D1029559

THE FRONTIERS COLLECTION

THE FRONTIERS COLLECTION

Series Editors:
A.C. Elitzur M.P. Silverman J. Tuszynski R. Vaas H.D. Zeh

The books in this collection are devoted to challenging and open problems at the forefront of modern science, including related philosophical debates. In contrast to typical research monographs, however, they strive to present their topics in a manner accessible also to scientifically literate non-specialists wishing to gain insight into the deeper implications and fascinating questions involved. Taken as a whole, the series reflects the need for a fundamental and interdisciplinary approach to modern science. Furthermore, it is intended to encourage active scientists in all areas to ponder over important and perhaps controversial issues beyond their own speciality. Extending from quantum physics and relativity to entropy, consciousness and complex systems – the Frontiers Collection will inspire readers to push back the frontiers of their own knowledge.

Information and Its Role in Nature
By J. G. Roederer

Relativity and the Nature of Spacetime
By V. Petkov

Quo Vadis Quantum Mechanics?
Edited by A. C. Elitzur, S. Dolev, N. Kolenda

Life – As a Matter of Fat
The Emerging Science of Lipidomics
By O. G. Mouritsen

Quantum–Classical Analogies
By D. Dragoman and M. Dragoman

Knowledge and the World
Challenges Beyond the Science Wars
Edited by M. Carrier, J. Roggenhofer, G. Küppers, P. Blanchard

Quantum–Classical Correspondence
By A. O. Bolivar

Mind, Matter and Quantum Mechanics
By H. Stapp

Quantum Mechanics and Gravity
By M. Sachs

Extreme Events in Nature and Society
Edited by S. Albeverio, V. Jentsch, H. Kantz

The Thermodynamic Machinery of Life
By M. Kurzynski

The Emerging Physics of Consciousness
Edited by J. A. Tuszynski

Weak Links
Stabilizers of Complex Systems from Proteins to Social Networks
By P. Csermely

Mind, Matter and the Implicate Order
By P.T.I. Pylkkänen

Quantum Mechanics at the Crossroads
New Perspectives from History, Philosophy and Physics
Edited by J. Evans, A.S. Thomdike

Particle Metaphysics
A Critical Account of Subatomic Reality
By B. Falkenburg

The Physical Basis of the Direction of Time
By H.D. Zeh

Asymmetry: The Foundation of Information
By S.J. Muller

Mindful Universe
Quantum Mechanics and the Participating Observer
By H. Stapp

Henry P. Stapp

MINDFUL UNIVERSE

Quantum Mechanics and the Participating Observer

With 9 Figures

Springer

Henry P. Stapp
University of California, Berkeley,
Lawrence Berkeley National Laboratory
email: hpstapp@lbl.gov

Cover figure: Image courtesy of the Scientific Computing and Imaging Institute, University of Utah (www.sci.utah.edu).

Library of Congress Control Number: 2007926114

ISSN 1612-3018
ISBN 978-3-540-72413-1 Springer Berlin Heidelberg New York

Springer is a part of Springer Science+Business Media

springer.com

Typesetting: Digital data supplied by Author
Production: LE-TEX Jelonek, Schmidt & Vöckler GbR, Leipzig
Cover design: KünkelLopka, Werbeagentur GmbH, Heidelberg

Printed on acid-free paper SPIN 12277445 57/3180/YL - 5 4 3 2 1

For Olivia

Preface

This book concerns your nature as a human being. It is about the connection of your mind to your body.

You may imagine that your mind – your stream of conscious thoughts, ideas, and feelings – influences your actions. You may believe that what you think affects what you do. You could be right. However, the scientific ideas that prevailed from the time of Isaac Newton to the beginning of the twentieth century proclaimed your physical actions to be completely determined by processes that are describable in physical terms alone. Any notion that your conscious choices make a difference in how you behave was branded an illusion: you were asserted to be causally equivalent to a mindless automaton.

We now know that that earlier form of science is fundamentally incorrect. During the first part of the twentieth century, that classical-physics-based conception of nature was replaced by a new theory that reproduces all of the successful predictions of its predecessor, while providing also valid predictions about a host of phenomena that are strictly incompatible with the precepts of eighteenth and nineteenth century physics. No prediction of the new theory has been shown to be false.

The new theory departs from the old one in many important ways, but none is more significant in the realm of human affairs than the role it assigns to your conscious choices. These choices are not fixed by the laws of the new physics, yet these choices are asserted by those laws to have important causal effects in the physical world. Thus contemporary physical theory annuls the claim of mechanical determinism. In a profound reversal of the classical physical principles, its laws make your conscious choices causally effective in the physical world, while failing to determine, even statistically, what those choices will be.

More than three quarters of a century have passed since the overturning of the classical laws, yet the notion of mechanical determinism still dominates the general intellectual milieu. The inertia of that superceded physical theory continues to affect your life in important ways. It still drives the decisions of governments, schools, courts, and medical institutions, and even your own choices, to the extent that you are influenced by what you are told by pundits who expound as scientific truth a mechanical idea of the universe that contravenes the precepts of contemporary physics.

The aim of this book is to explain to educated lay readers these twentieth century developments in science, and to touch upon the social consequences of the misrepresentations of contemporary scientific knowledge that continue to hold sway, particularly in the minds of our most highly educated and influential thinkers.

Acknowledgements

This work has benefited greatly from comments by K. Augustyn, R. Benin, J. Finkelstein, D. Lichtenberg, T. Nielsen, M. Velmans, T. Wallace, my wife Olivia, my son Henry, and especially from massive feedbacks from Edward Kelly and Adam Crabtree. Appendices D–G are contributions by me to a Compendium of Quantum Physics to be published by Springer, and the Atmanspacher interview in Chap. 15 was published in the September 2006 issue of the Journal of Consciousness Studies. I thank Jeffrey Schwartz for numerous suggestions pertaining to the form and content of this work.

Berkeley, *Henry P. Stapp*
February 2007

Contents

1 Science, Consciousness and Human Values

A tremendous burgeoning of interest in the problem of consciousness is now in progress. The grip of the behaviorists who sought to banish consciousness from science has finally been broken. This shift was ratified, for example, by the appearance several years ago of a special issue of Scientific American entitled *The Hidden Mind* (August 2002).

The lead article, written by Antonio Damasio, begins with the assertion: "At the start of the new millennium, it is apparent that one question towers above all others in the life sciences: How does the set of processes we call mind emerge from the activity of the organ we call brain?" He notes that some thinkers "believe the question to be unanswerable in principle", while: "For others, the relentless and exponential increase in knowledge may give rise to the vertiginous feeling that no problem can resist the assault of science *if only the science is right* and the techniques are powerful enough" (my emphasis). He notes that: "The naysayers argue that exhaustive compilation of all these data (of neuroscience) adds up to correlates of mental states but to nothing resembling *an actual mental state*" (his emphasis). He adds that: "In fact, the explanation of the physics related to biological events is still incomplete" and states that "the finest level of description of mind [...] might require explanation at the quantum level." Damasio makes his own position clear: "I contend that the biological processes now presumed to correspond to mind in fact are mind processes and will be seen to be so when understood in sufficient detail."

Damasio at least hints at the idea that "biological process [...] understood in sufficient detail" is a quantum understanding.

The possibility that quantum physics might be relevant to the connection between conscious process and brain process was raised also by Dave Chalmers, in his contribution 'The Puzzle of Conscious Experience' to *The Hidden Mind*. However, Chalmers effectively tied that possibility to the proposal put forth by Roger Penrose (1989, 1994) and, faulting that particular approach, rejected the general idea.

The deficiency of Penrose's approach identified by Chalmers is that it fails to bring in consciousness. It is about certain brain processes that may be related to consciousness, but "the theory is silent about how these processes might give rise to conscious experience. Indeed, the same problem arises with any theory of consciousness based only on physical processing."

Penrose's treatment does indeed focus on physical processing. But quantum theory itself is intrinsically psychophysical: as designed by its founders, and as used in actual scientific practice, it is ultimately a theory about the structure of our experience that is erected upon a radical mathematical generalization of the laws of classical physics.

Chalmers goes on to expound upon the 'explanatory gap' between, on the one hand, theoretical understanding of the behavioral and functional aspects of brain processes and, on the other hand, an explanation of how and why the performance of those functions should be accompanied by conscious experience. Such a gap arises in the classical approximation, but not in orthodox quantum theory, which is fundamentally a causal weaving together of the structure of our streams of conscious experiences, described in psychological terms, with a theoretical representation of the physical world described in mathematical language.

The conflating of Nature herself with the impoverished mechanical conception of it invented by scientists during the seventeenth century has derailed the philosophies of science and of mind for more than three centuries, by effectively eliminating the causal link between the psychological and physical aspects of nature that contemporary physics restores.

But the now-falsified classical conception of the world still exerts a blinding effect. For example, Daniel Dennett (1994, p. 237) says that his own thinking rests on the idea that "a brain was always going to do what it was caused to do by current, local, mechanical circumstances". But by making that judgment he tied his thinking to the physical half of Cartesian dualism, or its child, classical physics, and thus was forced in his book *Consciousness Explained* (Dennett 1991) to leave consciousness out, as he himself admits, and tries to justify, at the end of the book. By effectively restricting himself to the classical approximation, which squeezes the effects of consciousness out of the more accurate consciousness-dependent quantum dynamics, Dennett cuts himself off from any possibility of validly explaining the physical efficacy of our conscious efforts.

Francis Crick and Christof Koch begin their essay in *The Hidden Mind* entitled 'The Problem of Consciousness' with the assertion: "The overwhelming question in neurobiology today is the relationship between the mind and the brain." But after a brief survey of the difficulties in getting an answer they conclude that: "Radically new concepts may indeed be needed – recall the modifications in scientific thinking forced on us by quantum mechanics. The only sensible approach is to press the experimental attack until we are confronted with dilemmas that call for new ways of thinking."

However, the two cases compared by Crick and Koch are extremely dissimilar. The switch to quantum theory was forced upon us by the fact that we had a very simple system – consisting of a single hydrogen atom interacting with the electromagnetic field – that was so simple that it could be exactly solved by the methods of classical physics, but the calculated answer did not agree with the empirical results. There was initially no conceptual problem. It was rather that precise computations were possible, but gave wrong answers. Here the problem is reversed: precise calculations of the dynamical brain processes associated with conscious experiences are not yet possible, and hence have not revealed any mismatch between theory and experiment. The problem is, rather, a conceptual one: the concepts of classical physics that many neurobiologists are committed to using are logically inadequate because, unlike the concepts of quantum physics, they effectively exclude our conscious thoughts.

Dave Chalmers emphasizes this conceptual difficulty, and concludes that experimental work by neurobiologists is not by itself sufficient to resolve 'The Puzzle of Conscious Experience'. Better concepts are also needed. He suggests that the stuff of the universe might be information, but then, oddly, rejects the replacement of classical physical theory, which is based on material substance, by quantum theory, which is built on an informational structure that causally links experienced increments of knowledge to physically described processes.

During the nineteenth century, before the precepts of classical physics had been shown to be false at the fundamental level, scientists and philosophers had good reasons to believe that the physical aspects of reality were causally closed: that the mathematically described physical aspects of nature were completely determined, by the laws of Nature, in terms of earlier properties of the same kind. However, even then this led to a certain unreasonableness noted by William James (1890, p. 138): consciousness seems to be "an organ, superadded to the other organs which maintain the animal in its struggle for existence; and the presumption of course is that it helps him in some way in this struggle, just as they do. But it cannot help him without being in some way efficacious and influencing the course of his bodily history." James went on to examine the circumstances under which consciousness appears, and ended up saying: "The conclusion that it is useful is, after all this, quite justifiable. But if it is useful it must be so through its causal efficaciousness, and the automaton-theory must succumb to common-sense" (James 1890, p. 144).

That was James's conclusion even at a time when deterministic classical physical theory seemed secure and unchallengeable, and the notion that we human beings are mechanical automata was the rationally inescapable consequence of a triumphant physics. James's analysis was vindicated, however, by the ascendancy of quantum mechanics during the first half of the twentieth century. The aim of this book is to describe the development of this revised conceptualization of the connection between our minds and our brains, and the consequent revision of the role of human consciousness in the unfolding of reality. This revision in our understanding of ourselves and our place in nature infuses the subject with a significance that extends far beyond the narrowly construed boundaries of science. These changes penetrate to the heart of important sociological and philosophical issues.

Science has improved our lives in many ways. It has lightened the load of tedious tasks and expanded our physical powers, thereby contributing to a great flowering of human creativity. On the other hand, it has given us also the capacity to ravage the environment on an unprecedented scale and to obliterate our species altogether. Yet along with this fatal power it has provided a further offering which, though subtle in character and still hardly felt in the minds of men, may ultimately be its most valuable contribution to human civilization, and the key to human survival.

Science is not only the enterprise of harnessing nature to serve the practical needs of humankind. It is also part of man's unending search for knowledge about the universe and his place within it. This quest is motivated not solely by idle curiosity. Each of us, when trying to establish values upon which to base conduct, is inevitably led to the question of one's place in the greater whole. The linkage of this philosophical inquiry to the practical question of personal values is no mere intellectual abstraction. Martyrs in every age are vivid reminders of the fact that no influence upon human conduct, even the instinct for bodily self-preservation, is stronger than beliefs about one's relationship to the rest of the universe and to the power that shapes it. Such beliefs form the foundation of a person's self-image, and hence, ultimately, of personal values.

It is often claimed that science stands mute on questions of values: that science can help us to achieve what we value once our priorities are fixed, but can play no role in fixing these weightings. That claim is certainly incorrect. Science plays a key role in these matters. For what we value depends on what we believe, and what we believe is strongly influenced by science.

A striking example of this influence is the impact of science upon the system of values promulgated by the church during the Middle Ages. That structure rested on a credo about the nature of the universe, its creator, and man's connection to that creator. Science, by casting doubt upon that belief, undermined the system of values erected upon it. Moreover, it put forth a credo of its own. In that 'scientific' vision we human beings were converted from sparks of divine creative power, endowed with free will, to mechanical automata – to cogs in a giant machine that grinds inexorably along a preordained path in the grip of a blind causal process.

This material picture of human beings erodes not only the religious roots of moral values but the entire notion of personal responsibility. Each of us is asserted to be a mechanical extension of what existed prior to his or her birth. Over that earlier situation one has no control. Hence for what emerges, preordained, from that prior state one can bear no responsibility.

This conception of man undermines the foundation of rational moral philosophy, and science is doubly culpable: It not only erodes the foundations of earlier value systems, but also acts to strip man of any vision of himself and his place in the universe that could be the rational basis for an elevated set of values.

During the twentieth century this morally corrosive mechanical conception of nature was found to be profoundly incorrect. It failed not just in its fine details, but at its fundamental core. A vastly different conceptual framework was erected by the atomic physicists Werner Heisenberg, Niels Bohr, Wolfgang Pauli and their colleagues. Those scientists were forced to a wholesale revision of the entire subject matter of physical theory by the peculiar character of the new mathematical rules, which were invariably validated by reliable empirical data.

The earlier 'classical' physics had emerged from the study of the observed motions of the planets and large terrestrial objects, and the entire physical universe was, correspondingly, conceived to be made, essentially, out of miniaturized versions of these large visible objects.

Called "solid, massy, hard, impenetrable moveable particles" by Newton (1704), these tiny objects were conceived to act upon each other by contact interactions, much like billiard balls, except for the mysterious action at a distance called gravity. Newton himself rejected the idea that gravity could really act at distance without any intervening carrier. Nevertheless, provisional rules were found that were imagined to control the behavior of these tiny entities, and thus also the objects composed of them. These laws were independent of whether or not anyone was observing the physical universe: they took no special cognizance of any acts of observation performed by human beings, or of any knowledge acquired from such observations, or of the conscious thoughts of human beings. All such things were believed, during the reign of classical physics, to be completely determined, insofar as they had any physical consequences, by the physically described properties and laws that acted wholly mechanically at the microscopic scale. But the baffling features of new kinds of data acquired during the twentieth century caused the physicists who were studying these phenomena, and trying to ascertain the laws that governed them, to turn the whole scientific enterprise upside down.

Perhaps I should say that they turned right side up what had been upside down. For the word 'science' comes from the Latin word 'scire', 'to know', and what the founders of the new theory claimed, basically, is that the proper subject matter of science is not what may or may not be 'out there', unobserved and unknown to human beings. It is rather what we human beings can know, and can do in order to know more. Thus they formulated their new theory, called quantum mechanics, or quantum theory, around the knowledge-acquiring actions of human beings, and the knowledge we acquire by performing these actions, rather than around a conjectured causally sufficient mechanical world. The focus of the theory was shifted from one that basically ignored our knowledge to one that is about our knowledge, and about the effects of the actions that we take to acquire more knowledge upon what we are able to know.

This modified conception differs from the old one in many fascinating ways that continue to absorb the interest of physicists. However, it is the revised understanding of the nature of human beings, and of the causal role of human consciousness in the unfolding of reality, that is, I believe, the most exciting thing about the new physics, and probably, in the final analysis, also the most important contribution of science to the well-being of our species.

The rational foundation for this revised conceptual structure emerged from the intense intellectual struggles that took place during the twenties, principally between Niels Bohr, Werner Heisenberg, and Wolfgang Pauli. Those struggles replaced the then-prevailing Newtonian idea of matter as "solid, massy, hard, impenetrable, moveable particles" with a new concept that allowed, and in fact required, an entry into the causal structure of the physical effects of conscious decisions made by human subjects. This radical change swept away the meaningless billiard-ball universe, and replaced it with a universe in which we human beings, by means of our value-based intentional efforts, can make a difference first in our own behaviors, thence in the social matrix in which we are imbedded, and eventually in the entire physical reality that sustains our streams of conscious experiences.

The existing general descriptions of quantum theory emphasize puzzles and paradoxes in a way that tend to make non-physicists leery of using in any significant away the profound changes in our understanding of both man and nature wrought by the quantum revolution. Yet in the final analysis quantum mechanics is *more* understandable than classical mechanics because it is more deeply in line with our common sense ideas about our role in nature than the 'automaton' notion promulgated by classical physics. It is the three hundred years of indoctrination with mechanistic ideas that now makes puzzling a conception of ourselves that is fully concordant with both normal human intuition and the full range of empirical facts.

The founders of quantum mechanics presented this theory to their colleagues as essentially a set of rules about how to make predictions about the empirical feedbacks that we human observers will experience if we take certain actions. Classical mechanics can, of course, be viewed in exactly the same way, but the two theories differ profoundly in their logical and mathematical structures, and consequently, and even more profoundly, in what they purport to be fundamentally about.

In classical mechanics the state of any system, at some fixed time t, is defined by giving the location and the velocity of every particle in that system, and by giving also the analogous information about the electromagnetic and gravitational fields. All observers and their acts of observation are conceived to be simply parts or aspects of the continuously evolving fully mechanically pre-determined physically described universe. A person's stream of consciousness is considered to be some mysterious, but causally irrelevant or redundant, by-product or counterpart of his or her classically conceived and described brain activity.

But this classical idea that our conscious experiences are just some idea-like counterparts of a continuously evolving brain state encounters a certain difficulty. The classically conceived evolution of the brain is continuous, and hence the number of different physical states that occur during any temporal interval of continuous change is infinite. Thus a natural mind–brain connection should give, it would seem, a continuously changing state of consciousness, composed of parts in a way analogous to the neural activity that it represents. But this surmise seems at odds with the empirical evidence. According to William James (1911):

> [...] a discrete composition is what actually obtains in our perceptual experience. We either perceive nothing, or something already there in a sensible amount. This fact is what is known in psychology as the laws of the 'threshold'. Either your experience is of no content, of no change, or it is of a perceptible amount of content or change. Your acquaintance with reality grows literally by buds or drops of perception. Intellectually and on reflection you can divide these into components, but as immediately given they come totally or not at all.

A similar discreteness is the signature of quantum phenomena: the quantum wave is spread out over a vast region covering many detectors, but only one detector fires, the rest do not. The element of discreteness, the 'Yes' or 'No' of the Geiger counter's 'click' is an elemental feature of quantum theory. Thus Bohr (1962, p. 60) speaks of: "The element of wholeness, symbolized by the quantum of action and completely foreign to classical physical principles."

In psychology the identity and form of the percept that actually enters into a stream of consciousness depends strongly on the intention of the probing mind: a person tends to experience what he or she is looking for, provided the potentiality for that experience is present. The observer does not create what is not potentially there, but does participate in the extraction from the mass of existing potentialities individual items that have interest and meaning to the perceiving self.

Quantum theory exhibits, as we shall see, a similar feature. Thus both psychology and physics, when examined in depth, reveal observer-influenced whole elements that seem "foreign to classical physical principles".

Insofar as it has been tested, the new theory, quantum theory, accounts for all the observed successes of the earlier physical theories, and also for the immense accumulation of new data that the earlier concepts cannot accommodate. But, according to the new conception, the *physically described world* is built not out of bits of matter, as matter was understood in the nineteenth century, but out of objective *tendencies* – potentialities – for certain discrete, whole *actual events* to occur. Each such event has both a psychologically described aspect, which is essentially an increment in knowledge, and also a physically described aspect, which is an action that *abruptly changes* the mathematically described set of potentialities to one that is concordant with the increase in knowledge. This coordination of the aspects of the theory that are described in physical/mathematical terms with aspects that are described in psychological terms is what makes the theory practically useful. Some empirical predictions have been verified to the incredible accuracy of one part in a hundred million.

The most radical change wrought by this switch to quantum mechanics is the injection directly into the dynamics *of certain choices made by human beings about how they will act*. Human actions enter, of course, also in classical physics. But the two cases are fundamentally different. In the classical case the way a person acts is fully determined in principle by the physically described aspects of reality alone. But in the quantum case there is *an essential gap in physical causation*. This gap is generated by Heisenberg's uncertainty principle, which opens up, at the level of human actions, a range of alternative possible behaviors between which the physically described aspects of theory are in principle unable to choose or decide. But this loss-in-principle of causal definiteness, associated with a loss of knowable-in-principle physically describable information, opens the way, logically, to an input into the dynamics of another kind of possible causes, which are eminently knowable, both in principle and in practice, namely our conscious choices about how we will act. These interventions in the dynamics take the form of specifications of *new boundary conditions*.

The specifications of boundary conditions is, of course, the traditional job of the experimenters. But in classical physics the only needed setting of boundary conditions is the one done by God at the beginning of time. On the other hand, the conventional laws of quantum mechanics have both a dynamical opening for, and a logical need for, additional choices made later on. Thus contemporary orthodox physics delegates some of the responsibilities formerly assigned to an inscrutable God, acting in the distant past, to our present knowable conscious actions.

Niels Bohr emphasized this freedom of action of the experimenters in passages such as:

> The freedom of experimentation, presupposed in classical physics, is of course retained and corresponds to the free choice of experimental arrangement for which the mathematical structure of the quantum mechanical formalism offers the appropriate latitude. (Bohr 1958, p. 73)

> To my mind, there is no other alternative than to admit that, in this field of experience, we are dealing with individual phenomena and that our possibilities of handling the measuring instruments allow us only to make a choice between the different complementary types of phenomena that we want to study. (Bohr 1958, p. 51)

In John von Neumann's rigorous mathematical formulation of quantum mechanics the effects of these free choices upon the physically described world are specifically called 'interventions' (von Neumann 1955/1932, pp. 358, 418). These choices are 'free' in the sense that they are not coerced, fixed, or determined by the physically described aspects of the theory. Yet these choices, which are not fixed or determined by any law of orthodox contemporary physics, and which *seem to us* to depend partly upon 'reasons' based on felt values, definitely have potent effects upon the physically described aspects of the theory. These effects are specifically described by the theory.

Nothing like this effective action of mind upon physically described things exists in classical physics. There is nothing in the principles of classical physics that requires, or even hints at, the existence of such things as thoughts, ideas, and feelings, and certainly no opening for aspects of nature not determined by the physically describable aspects of nature to 'intervene' and thereby influence the future physically described structure. In fact, it is precisely the absence from classical physics of any notion of experiential-type realities, or of any job for them to do, or of any possibility for them to do anything not already done locally by the mechanical elements, that has been the bane of philosophy for three hundred years. Eliminating this scientifically unsupported precept of the causal closure of the physical opens the way to a new phase of science-based philosophy.

The preceding remarks give a brief overview of the theme of this work. I shall begin my more detailed account of these twentieth century developments in science by emphasizing, in the words of the founders themselves, the central role played in the new theory by 'our knowledge'.

2 Human Knowledge as the Foundation of Science

In the introduction to his book *Quantum Theory and Reality* the philosopher of science Mario Bunge (1967, p. 4) said:

> The physicist of the latest generation is operationalist all right, but usually he does not know, and refuses to believe, that the original Copenhagen interpretation – which he thinks he supports – was squarely subjectivist, i.e., nonphysical.

Let there be no doubt about this point. The original form of quantum theory is subjective, in the sense that it is forthrightly about relationships among conscious human experiences, and it expressly recommends to scientists that they resist the temptation to try to understand the reality responsible for the correlations between our experiences that the theory correctly describes. The following brief collection of quotations by the founders gives a conspectus of the Copenhagen philosophy:

> The conception of objective reality of the elementary particles has thus evaporated not into the cloud of some obscure new reality concept but into the transparent clarity of a mathematics that represents no longer the behavior of particles but rather our knowledge of this behavior. (Heisenberg 1958a, p. 100)

> [...] the act of registration of the result in the mind of the observer. The discontinuous change in the probability function [...] takes place with the act of registration, because it is the discontinuous change in our knowledge in the instant of registration that has its image in the discontinuous change of the probability function. (Heisenberg 1958b, p. 55)

> When the old adage "Natura non facit saltus" (Nature makes no jumps) is used as a basis of a criticism of quantum theory, we can reply that certainly our knowledge can change suddenly, and that this fact justifies the use of the term 'quantum jump'. (Heisenberg 1958b, p. 54)

> It was not possible to formulate the laws of quantum mechanics in a fully consistent way without reference to the consciousness. (Wigner 1961b, p. 169)

> In our description of nature the purpose is not to disclose the real essence of phenomena but only to track down as far as possible relations between the multifold aspects of our experience. (Bohr 1934, p. 18)

> Strictly speaking, the mathematical formalism of quantum mechanics merely offers rules of calculation for the deduction of expectations about observations obtained under well-defined classical concepts. (Bohr 1963, p. 60)

> [...] the appropriate physical interpretation of the symbolic quantum mechanical formalism amounts only to prediction of determinate or statistical character, pertaining to individual phenomena appearing under conditions defined by classical physics concepts. (Bohr 1958, p. 64)

The references to 'classical (physics) concepts' is explained by Bohr as follows:

> [...] it is imperative to realize that in every account of physical experience one must describe both experimental conditions and observations by the same means of communication as the one used in classical physics. Bohr (1958, p. 88)

> [...] we must recognize above all that, even when phenomena transcend the scope of classical physical theories, the account of the experimental arrangement and the recording of observations must be given in plain language supplemented by technical physical terminology. (Bohr 1958)

Bohr is saying that scientists do in fact use, and must use, the concepts of classical physics in communicating to their colleagues the specifications on how the experiment is to be set up, and what will constitute a certain type of outcome. He in no way claims or admits that there is an actual objective reality out there that conforms to the precepts of classical physics.

In his book *The Creation of Quantum Mechanics and the Bohr–Pauli Dialogue*, the historian John Hendry (1984) gives a detailed account of the fierce struggles by such eminent thinkers as Hilbert, Jordan, Weyl, von Neumann, Born, Einstein, Sommerfeld, Pauli, Heisenberg, Schroedinger, Dirac, Bohr and others, to come up with a rational way of comprehending the data from atomic experiments. Each man had his own bias and intuitions, but in spite of intense effort no rational comprehension was forthcoming. Finally, at the 1927 Solvay conference a group including Bohr, Heisenberg, Pauli, Dirac, and Born come into concordance on a solution that came to be called the Copenhagen interpretation, due to the central role of Bohr and those working with him at his institute in Denmark.

Hendry says: "Dirac, in discussion, insisted on the restriction of the theory's application to our knowledge of a system, and on its lack of ontological content." Hendry summarized the concordance by saying: "On this interpretation it was agreed that, as Dirac explained, the wave function represented our knowledge of the system, and the reduced wave packets our more precise knowledge after measurement."

These quotations make it clear that, in direct contrast to the ideas of classical physical theory, orthodox Copenhagen quantum theory is about 'our knowledge'. We, and in particular our mental aspects, have entered into the structure of basic physical theory.

This profound shift in physicists' conception of the basic nature of their endeavor, and of the meanings of their formulas, was not a frivolous move: it was a last resort. The very idea that in order to comprehend atomic phenomena one must abandon physical ontology, and construe the mathematical formulas to be directly about the knowledge of human observers, rather than about external reality itself, is so seemingly preposterous that no group of eminent and renowned scientists would ever embrace it except as an extreme last measure. Consequently, it would be frivolous of us simply to ignore a conclusion so hard won and profound, and of such apparent direct bearing on our effort to understand the connection of our conscious thoughts to our bodily actions.

Einstein never accepted the Copenhagen interpretation. He said:

> What does not satisfy me, from the standpoint of principle, is its attitude toward what seems to me to be the programmatic aim of all physics: the complete description of any (individual) real situation (as it supposedly exists irrespective of any act of observation or substantiation). (Einstein 1951, p. 667; the parenthetical word and phrase are part of Einstein's statement.)

and

> What I dislike in this kind of argumentation is the basic positivistic attitude, which from my view is untenable, and which

> seems to me to come to the same thing as Berkeley's principle, esse est percipi. [Transl: To be is to be perceived] (Einstein 1951, p. 669)

Einstein struggled until the end of his life to get the observer's knowledge back out of physics. He did not succeed! Rather he admitted (ibid. p. 87) that:

> It is my opinion that the contemporary quantum theory constitutes an optimum formulation of the [statistical] connections.

He also referred (ibid, p. 81) to:

> [...] the most successful physical theory of our period, viz., the statistical quantum theory which, about twenty-five years ago took on a logically consistent form. This is the only theory at present which permits a unitary grasp of experiences concerning the quantum character of micro-mechanical events.

One can adopt the cavalier attitude that these profound difficulties with the classical conception of nature are just some temporary retrograde aberration in the forward march of science. One may imagine, as some do, that a strange confusion has confounded our best minds for seven decades, and that the weird conclusions of physicists can be ignored because they do not fit a tradition that worked for two centuries. Or one can try to claim that these problems concern only atoms and molecules, but not the big things built out of them. In this connection Einstein said (ibid, p. 674): "But the 'macroscopic' and 'microscopic' are so inter-related that it appears impracticable to give up this program [of basing physics on the 'real'] in the 'microscopic' domain alone."

These quotations document the fact that Copenhagen quantum theory brings human consciousness into physical theory in an essential way. But how does this radical change in basic physics affect science's conception of the human person?

To answer this query I begin with a few remarks on the development of quantum theory.

The original version of quantum theory, called the Copenhagen quantum theory, or the Copenhagen interpretation, is forthrightly pragmatic. It aims to show how the mathematical structure of the theory can be employed to make useful, testable predictions about our future possible experiences on the basis of our past experiences and the forms of the actions that we choose to make. In this initial version of the theory the brains and bodies of the experimenters, and also their measuring devices, are described fundamentally in empirical terms: in terms of our experiences/perceptions pertaining to these devices and their manipulations by our physical bodies. The devices are treated as extensions of our bodies. However, the boundary between our empirically described selves and the physically described system we are studying is somewhat arbitrary. The empirically described measuring devices can become very tiny, and physically described systems can become very large, This ambiguity was examined by von Neumann (1932) who showed that we can consistently describe the entire physical world, including the brains of the experimenters, as the physically described world, with the actions instigated by an experimenter's stream of consciousness acting directly upon that experimenter's brain. The interaction between the psychologically and physically described aspects in quantum theory thereby becomes the mind–brain interaction of neuroscience and neuropsychology.

It is this von Neumann extension of Copenhagen quantum theory that provides the foundation for a rationally coherent ontological interpretation of quantum theory – for a putative description of what is really happening. Heisenberg suggested an ontological description in his 1958 book *Physics and Philosophy* and I shall adhere to that ontology, formulated within von Neumann's framework in which the brain, as part of the physical world, is described in terms of the quantum mathematics. This localizes the mind–matter problem at the interface between the quantum mechanically described brain and the experientially described stream of consciousness of the human agent/observer.

My aim in this book is to explain to non-physicist the interplay between the psychologically and physically described components of mind–brain dynamics, as it is understood within the orthodox (von Neumann–Heisenberg) quantum framework.

3 Actions, Knowledge, and Information

3.1 The Anti-Newtonian Revolution

From the time of Isaac Newton until about 1925 science relegated consciousness to the role of passive viewer: our thoughts, ideas, and feelings were treated as impotent bystanders to a march of events wholly controlled by microscopically describable interactions between mechanically behaving microscopic basic elements. The founders of quantum mechanics made the revolutionary move of bringing conscious human experiences into basic physical theory in a fundamental way. After two hundred years of neglect, our thoughts were suddenly thrust into the limelight. This was an astonishing reversal of precedent because the enormous successes of the prior physics were due in large measure to the policy of excluding all mention of idea-like qualities from the formulation of the physical laws.

What sort of crisis could have forced the creators of quantum theory to contemplate, and eventually embrace, this radical idea of injecting our thoughts explicitly into the basic laws of physics?

The answer to this question begins with a discovery that occurred at the end of the nineteenth century. In December of 1900 Max Planck announced the discovery and measurement of the 'quantum of action'. Its measured value is called Planck's constant. This constant specifies one of three basic quantities that are built into the fundamental fabric of the physical universe. The other two are the gravitational constant, which fixes the strength of the force that pulls every bit of matter in the solar system toward every other bit, and the speed of light, which controls the response of every particle to this force, and to every other force. The integration into physics of each of these three basic quantities generated a monumental shift in our conception of nature.

Isaac Newton discovered the gravitational constant, which linked our understandings of celestial and terrestrial dynamics. It connected the motions of the planets and their moons to the trajectories of cannon balls here on earth, and to the rising and falling of the tides. Insofar as his laws are complete the entire physical universe is governed by mathematical equations that link every bit of matter to every other bit, and moreover fix the complete course of history for all times from physical conditions prevailing in the primordial past.

Einstein recognized that the 'speed of light' is not just the rate of propagation of some special kind of wave-like disturbance, namely 'light'. It is rather a fundamental number that enters into the equations of motion of every kind of material substance, and, among other things, prevents any piece of matter from traveling faster than this universal maximum value. Like Newton's gravitational constant it is a number that enters ubiquitously into the basic structure of Nature.

But important as the effects of these two quantities are, they are, in terms of profundity, like child's play compared to the consequences of Planck's discovery.

Planck's 'quantum of action' revealed itself first in the study of light, or, more generally, of electromagnetic radiation. The radiant energy emerging from a tiny hole in a heated hollow container can be decomposed into its various frequency components. Classical nineteenth century physics gave a prediction about how that energy should be distributed among the frequencies, but the empirical facts did not fit that theory. Eventually, Planck discovered that the empirically correct formula could be obtained by assuming essentially that the energy was concentrated in finite packets, with the amount of energy in each such unit being directly proportional to the frequency of the radiation that was carrying it. The ratio of energy to frequency is called Planck's constant. Its value is extremely small on the scale of normal human activity, but becomes significant when we come to the behavior of the atomic particles and fields out of which our bodies, brains, and the large physical objects around us are made.

Planck's discovery shattered the classical laws that had been for two centuries the foundation of the scientific world view. During the years that followed many experiments were performed on systems whose behaviors depend sensitively upon the properties of their atomic constituents. It was repeatedly found that the classical principles did not work: they gave well defined predictions that turned out to be flat-out wrong, when confronted with the experimental evidence. The fundamental laws of physics, which every physics student had been taught, and upon which much of the industrial and technological world of that era was based, were failing. More importantly, and surprisingly, they were failing in ways that no mere tinkering could ever fix. Something was fundamentally amiss. No one could say how these laws, which were so important, and that had seemed so perfect, could be fixed. No one could foresee whether a new theory could be constructed that would explain these strange and unexpected results, and restore rational order to our understanding of nature. But one thing was clear to those working feverishly on the problem: Planck's constant was somehow at the center of it all.

3.2 The World of Actions

Werner Heisenberg was, from a technical point of view, the principal founder of quantum theory. He discovered in 1925 the completely amazing and wholly unprecedented solution to the puzzle: the quantities that classical physical theory was based upon, and which were thought to be numbers, must be treated not as numbers but as actions! Ordinary numbers, such as 2 and 3, have the property that the product of any two of them does not depend on the order of the factors: 2 times 3 is the same as 3 times 2. But Heisenberg discovered that one could get the correct answers out of the old classical laws if one decreed that certain numbers that occur in classical physics as the magnitudes of certain physical properties of a material system are not ordinary numbers. Rather, they must be treated as *actions* having the property that the order in which they act matters!

This 'solution' may sound absurd or insane. But mathematicians had already discovered that logically consistent generalizations of ordinary mathematics exist in which numbers are replaced by 'actions' having the property that the order in which they are applied matters. The ordinary numbers that we use for everyday purposes like buying a loaf of bread or paying taxes are just a very special case from among a broad set of rationally coherent mathematical possibilities. In this simplest case, A times B happens to be the same as B times A. But there is no logical reason why Nature should not exploit one of the more general cases: there is no compelling reason why our physical theories must be based exclusively on ordinary numbers rather than on actions. The theory based on Heisenberg's discovery exploits the more general logical possibility. It is called quantum mechanics, or quantum theory.

The difference between quantum mechanics and classical mechanics is specified by Planck's constant, which is a tiny number on the scale of human actions. Thus this tweaking of laws of physics might seem to be a bit of mathematical minutia that could scarcely have any great bearing on the fundamental nature of the universe, or of our role within it. But replacing *numbers* by *actions* upsets the whole apple cart. It produced a seismic shift in our ideas about both the nature of reality, and the nature of our relationship to the reality that envelops and sustains us. The aspects of nature represented by the theory are converted from elements of *being* to elements of *doing*. The effect of this change is profound: it replaces the world of *material substances* by a world populated by *actions*, and by *potentialities* for the occurrence of the various possible observed feedbacks from these actions. Thus this switch from 'being' to 'action' allows – and according to orthodox quantum theory demands – a draconian shift in the very subject matter of physical theory, from an imagined universe consisting of causally self-sufficient mindless matter, to a universe populated by allowed possible physical actions and possible experienced feedbacks from such actions. A purported theory of matter alone is converted into a theory of the relationship between matter and mind.

What is this momentous change introduced by Heisenberg?

In classical physics the center point of each physical object has, at each instant of time, a well defined location, which can be specified by giving its three coordinates (x, y, z) relative to some coordinate system. For example, the location of (the center point of) a spider dangling in a room can be specified by letting z be its distance from the floor, and letting x and y be its distances from two intersecting walls. Similarly, the velocity of that dangling spider, as she drops to the floor, blown by a gust of wind, can be specified by giving the rates of change of these three coordinates (x, y, z). If each of these three *rates of change*, which together specify the velocity, are multiplied by the *weight* (= mass) of the spider, then one gets three numbers, say (p, q, r), that define the *momentum* of the spider. In classical physics one uses the set of three numbers denoted by (x, y, z) to represent the position of the center point of an object, and the set of three numbers labeled by (p, q, r) to represent the momentum of that object. These six numbers are just ordinary numbers that obey the commutative property of multiplication that we all, hopefully, learned in third grade: $x * p$ equals $p * x$, where $*$ means multiply.

The six-dimensional space of all possible values $(x, y, z; p, q, r)$ is called *phase space*: it is the space of all possible instantaneous 'states' of the particle.

Heisenberg's analysis showed that in order to make the formulas of classical physics work in general, $x * p$ must be different from $p * x$. He found that the difference between these two products must be Planck's constant. (Actually, the difference is Planck's constant divided by 2π and multiplied by the imaginary unit i, which is a number such that i times i is minus one.) Thus modern quantum theory was born by recognizing, or declaring, that the symbols used in classical physical theory to represent ordinary numbers actually represent actions such that their ordering in a sequence of actions matters. The procedure of creating the mathematical structure of quantum mechanics from that of classical physics, by replacing numbers by corresponding actions, is called 'quantization'.

The idea of replacing the numbers that specify where a particle is, and how fast it is moving, by mathematical quantities that violate the simple laws of arithmetic may strike you – if this is the first you've heard about it – as a giant step in the wrong direction. You might mutter that scientists should try to make things simpler, rather than abandoning one of the things we really know for sure, namely that the order in which one multiplies factors does not matter. But against that intuition one must recognize that this change works beautifully in practice: all of the tested predictions of quantum mechanics are borne out, and these include predictions that are correct to the incredible accuracy of one part in a hundred million. There must be something very, very right about this replacement of numbers by actions.

In classical physical theory each elementary particle is asserted to have at each instant of time a definite location, defined by a set of three numbers (x, y, z), and definite momentum, defined by a set of three numbers (p, q, r). In quantum theory one generally considers systems of many particles, but insofar as one can consider one particle alone the state of that particle at any instant of time would be represented by a *cloud of pairs of numbers*, with one pair of numbers (called a complex number) assigned to each point in three-dimensional (position) space. Someone might choose to perform a phenomenologically (i.e., experimentally/experientially) described probing action on this 'particle'. In quantum mechanics each such possible probing action turns out to have an associated set of distinct *experientially distinguishable* possible outcomes. The cloud of numbers *taken as a whole* determines the probability for the appearance of each of the alternative possible outcomes of that chosen probing action. The theory thus gives specified rules for computing the probabilities for each of the distinct alternative possible empirically described feedbacks from each of the alternative possible experimental probing actions that the human experimenter might chose to perform, but no rules that specify which probing action he or she will choose.

In classical physical theory when one descends from the macroscopic world of visible objects to the microscopic world of their elementary constituents one arrives at a world containing the 'solid, massy, hard, impenetrable moveable particles' that Newton spoke of. But in quantum theory one arrives instead at clouds, or quantum smears, of numbers that *taken as a whole* have empirical meaning in terms of probabilities of alternative possible experiences.

Briefly stated, the orthodox formulation of quantum theory (see Appendix D) asserts that, in order to connect adequately the mathematically described state of a physical system to human experience, there must be an abrupt *intervention* in the otherwise smoothly evolving mathematically described state of that system.

According to the orthodox formulation, these interventions are probing actions *instigated by human agents who are able to 'freely' choose which one, from among various alternative possible probing actions, they will perform.* The physically describable effect of the chosen probing action is to separate (partition) the prior physical state of the system being probed in some particular way into a set of component parts. Each physically described part corresponds to one *perceivable* outcome from the set of distinct alternative possible perceivable outcomes of that particular probing action.

If such a probing action is performed, then *one* of its allowed perceivable feedbacks will appear in the stream of consciousness of the observer, and the mathematically described state of the probed system will then jump abruptly from the form it had prior to the intervention to the partitioned portion of that state that corresponds to the observed feedback. *This means that, according to orthodox contemporary physical theory, the 'free' choices of probing actions made by agents enter importantly into the course of the ensuing psychologically and physically described events. Here the word 'free' means, however, merely that the choice is not determined by the (currently) known laws of physics; not that the choice has no cause at all in the full psychophysical structure of reality. Presumably the choice has some cause or reason – it is unreasonable that it should simply pop out of nothing at all – but the existing theory gives no reason to believe that this cause must be determined exclusively by the physically described aspects of the psychophysically described nature alone.*

If one sets Planck's constant equal to zero in the quantum mechanical equations then one recovers (the fundamentally incorrect) classical mechanics. Thus classical physics is *an approximation* to quantum physics. It is the approximation in which Planck's constant, wherever it appears, is replaced by zero. In this approximation the quantum smearing does not occur – each cloud is reduced to a point – and one recovers classical physics, along with the physical determinism (the causal closure of the physical) entailed by classical physics.

In the classical approximation there is no need for, *and indeed no room for*, any effect of any probing action. The *uncertainty* – arising from the non-zero size of the quantum cloud – that in the unapproximated theory needs to be resolved by the intervention of some particular probing action is already reduced to zero by the replacement of Planck's constant by zero. Thus all effects upon the physically/mathematically described aspects of nature's process that are instigated by the actions 'freely' chosen by agents are eliminated by the classical approximation. *Consequently*, any attempt to understand or explain within the framework of classical physics the physical effects of consciousness is irrational, *because the classical approximation eliminates the effect one is trying to study.*

3.3 Intentional Actions and Experienced Feedbacks

The concept of intentional actions by agents is of central importance. Each such action is intended to produce an experiential feedback. For example, a scientist might act to place a Geiger counter near a radioactive source, with the intention to see the counter either 'fire', or 'not fire', during a certain time interval. The experienced response, 'Yes' or 'No', to the query 'Does the counter fire?' specifies one bit of information. The basic move in quantum theory is to shift, *fundamentally*, from the airy plane of high-level abstractions, such as the unseen precise trajectories of invisible elementary material particles, to the nitty-gritty realities of consciously chosen intentional actions and their experienced feedbacks, and to the theoretical specification of the mathematical procedures that allow us successfully to predict relationships among these empirical realities.

Probing actions of this kind are performed not only by scientists. Every healthy and alert infant is engaged in making willful efforts that produce experiential feedbacks, and he or she soon begins to form expectations about what sorts of feedbacks are likely to follow from some particular kind of felt effort. Thus both empirical science and normal human life are based on paired realities of this action–response kind, and our physical and psychological theories are both basically attempts to understand these linked realities within a rational conceptual framework.

A purposeful action by a human agent has two aspects. One aspect is his conscious intention, which is described in psychological terms. The other aspect is the linked physical action, which is described in physical terms; i.e., in terms of mathematical entities assigned to spacetime points. For successful living the physically described action should be a *functional counterpart* of the conscious intention: after sufficient empirical honing by effective learning processes the physically described aspect of the felt intentional act should have a tendency to produce the intended experiential feedback.

John von Neumann, in his seminal book, *Mathematical Foundations of Quantum Mechanics*, calls by the name 'process 1' the basic probing action that partitions a potential continuum of physically described possibilities into a (countable) set of empirically recognizable alternative possibilities. I shall retain that terminology. Von Neumann calls the orderly mechanically controlled evolution that occurs between interventions by name 'process 2'. This process is the one controlled by the Schroedinger equation. The numbering, 1 and 2, emphasizes the important fact that the conceptual framework of orthodox quantum theory requires *first* an acquisition of knowledge, and *second*, a mathematically described propagation of a representation of this acquired knowledge to some later time at which a further inquiry is made.

There are two other associated processes that need to be recognized. The first of these is the process that *selects the outcome*, 'Yes' or 'No', of the probing action. Dirac calls this intervention a "choice on the part of nature", and it is subject, according to quantum theory, to statistical rules specified by the theory. I call by the name 'process 3' this statistically specified choice of the *outcome* of the action selected by the prior process 1 probing action

Finally, in connection with each process 1 action, there is, presumably, some process that is not described by contemporary quantum theory, but that determines what the so-called 'free choice' of the experimenter will actually be. This choice *seems to us* to arise, at least in part, from conscious reasons and valuations, and it is certainly strongly influenced by the state of the brain of the experimenter. I have previously called this selection process by the name 'process 4', but will use here the more apt name 'process zero', because this process must precede von Neumann's process 1. It is the absence from orthodox quantum theory of any description on the workings of process zero that constitutes the causal gap in contemporary orthodox physical theory. It is this 'latitude' offered by the quantum formalism, in connection with the "freedom of experimentation" (Bohr 1958, p. 73), that blocks the causal closure of the physical, and thereby releases human actions from the immediate bondage of the physically described aspects of reality.

3.4 Cloudlike Forms

The quantum state of a single elementary particle can be visualized, roughly, as a continuous cloud of (complex) numbers, one assigned to every point in three-dimensional space. This cloud of numbers evolves in time and, taken as a whole, it determines, at each instant, for each allowed process 1 action, an associated set of alternative possible experiential outcomes or feedbacks, and the 'probability of finding (i.e., experiencing)' that particular outcome.

Heisenberg's uncertainty principle specifies that if one squeezes this spatial cloud – the spatial region in which the numbers are nonzero – into a sufficiently small region, it will violently explode outward when the constricting force is removed.

3.5 Simple Harmonic Oscillators

One of the most important and illuminating examples of this cloudlike feature of the quantum state is the one corresponding to a pendulum, or more precisely, to what is called a simple harmonic oscillator. Such a system is one in which there is a restoring force that tends to push the center point of the object to a single 'base point', and in which the strength of this restoring force is directly proportional to the distance of the center point of the object from this base point.

According to classical physics any such system has a state of lowest possible energy. In this state the center point of the object lies motionless at the base point. In quantum theory this system again has a state of lowest possible energy. But this state is not localized at the base point. It is a cloudlike spatial structure that is spread out over a region that extends to infinity. However, the probability distribution represented by this cloudlike form has the shape of a bell: it is largest at the base point, and falls off in a prescribed manner as the distance of the center point from the base point increases.

If one were to put this state of lowest energy into a container, then squeeze it into a more narrow space, and then let it loose, the cloudlike form would explode outward, but then settle into an oscillating motion. Thus the cloudlike spatial structure behaves rather like a swarm of bees, such that the more they are squeezed in space the faster they move relative to their neighbors, and the faster the squeezed cloud will explode outward if the squeezing constraint is released. This 'explosive' property of narrowly confined states plays a key role in quantum brain dynamics, as we shall soon see. This explosive property is a consequence of Heisenberg's uncertainty principle, which entails that a severe confinement of the cloud in ordinary (coordinate) space entails a large spread in a corresponding cloud in momentum (hence velocity) space.

3.6 The Double-Slit Experiment

There is a crucial difference between the behavior of the quantum cloudlike form and the somewhat analogous probability distribution of classical statistical mechanics. This difference is exhibited by the famous double-slit experiment. If one shoots an electron, a calcium ion, or any other quantum counterpart of a tiny classical object, at a narrow slit then if the object passes through the slit the associated cloudlike form will fan out over a wide angle, due essentially to the reaction to squeezing mentioned above. But if one opens two closely neighboring narrow slits, then what passes through the slits is described by a probability distribution that is not just the sum of the two separate fanlike structures that would be present if each slit were opened separately. Instead, at some points the probability value will be nearly twice the *sum* of the values associated with the two individual slits, and in other places the probability value drops nearly to zero, even though both individual fanlike structures give a large probability value at that place. This non-additivity – or interference – property of the quantum cloudlike structure makes that structure very different from a probability distribution of classical physics, because in the classical case the probabilities arising from the two individual slits will simply add.

This non-additivity property, which holds for a quantum particle such as an electron or a calcium ion, persists even when the particles come one at a time! According to *classical* ideas each tiny individual object must pass through either one slit or the other, so the probability distribution must be just the sum of the contributions from the two separate slits. But this is not what happens empirically. Quantum mechanics deals consistently with this non-additivity property, and with all the other non-classical properties of these cloudlike structures. The non-additivity property is not at all mysterious or strange if one accepts the basic idea that reality is not made out of any material substance, but rather out of 'events' (actions) and 'potentialities' for these events to occur. Potentialities are not material realities, and there is no logical requirement that they be simply additive. According to the mathematically consistent rules of quantum theory, the quantum potentialities are not simply additive: they have a wave-like nature, and can interfere like waves.

4 Nerve Terminals and the Need to Use Quantum Theory

Many neuroscientists who study the relationship of consciousness to brain processes want to believe that classical physics will provide an adequate rational foundation for that task. But classical physics has bottom-up causation, and the direct rational basis for the claim that classical physics is applicable to the full workings of the brain rests on the basic presumption that it is applicable at the microscopic level. However, empirical evidence about what is actually happening at the trillions of synapses on the billions of neurons in a conscious brain is virtually nonexistent, and, according to the uncertainty principle, empirical evidence is *in principle* unable to justify the claim that deterministic behavior actually holds in the brain at the microscopic (ionic) scale. Thus the claim that classical determinism holds in living brains is empirically indefensible: sufficient evidence neither does, nor can in principle, exist.

Whether the classical approximation is applicable to macroscopic brain dynamics can, therefore, only be determined by examining the details of the physical situation within the framework of the more general quantum theory, to see, from a rational perspective, to what extent use of the classical approximation can be theoretically justified. The technical questions are: How important *quantitatively* are the effects of the uncertainty principle at the microscopic (ionic) level; and if they are important at the microscopic level, then why can this microscopic indeterminacy never propagate up to the macro-level?

Classical physical theory is adequate, in principle, precisely to the extent that the smear of potentialities generated at the microscopic level by the uncertainty principle leads via the purely physically described aspects of quantum dynamics to a macroscopic brain state that is essentially one single classically describable state, rather than a cloud of such states representing a set of *alternative* possible conscious experiences. In this latter case the quantum mechanical state of the brain needs to be *reduced*, somehow, to the state corresponding to the experienced phenomenal reality.

To answer the physics question of the extent of the micro-level uncertainties we turn first to an examination of the quantum dynamics of nerve terminals.

4.1 Nerve Terminals

Nerve terminals lie at the junctions between two neurons, and mediate the functional connection between them. Neuroscientists have developed, on the basis of empirical data, fairly detailed classical models of how these important parts of the brain work. According to the classical picture, each 'firing' of a neuron sends an electrical signal, called an action potential, along its output fiber. When this signal reaches the nerve terminal it opens up tiny channels in the terminal membrane, through which calcium ions flow into the interior of the terminal. Within the terminal are vesicles, which are small storage areas containing chemicals called neurotransmitters. The calcium ions migrate by diffusion from their entry channels to special sites, where they trigger the release of the contents of a vesicle into a gap between the terminal and a neighboring neuron. The released chemicals influence the tendency of the neighboring neuron to fire. Thus the nerve terminals, as connecting links between neurons, are basic elements in brain dynamics.

The channels through which the calcium ions enter the nerve terminal are called ion channels. At their narrowest points they are only about a nanometer in width, hence not much larger than the calcium ions themselves. This extreme smallness of the opening in the ion channels has profound quantum mechanical import. The consequence of this narrowness is essentially the same as the consequence of the squeezing of the state of the simple harmonic oscillator, or of the narrowness of the slits in the double-slit experiments. The narrowness of the channel restricts the lateral spatial dimension. Consequently, the uncertainty in lateral velocity is forced by the quantum uncertainty principle to become non-zero, and to be in fact about 1% of the longitudinal velocity of the ion. This causes the quantum probability cloud associated with the calcium ion to fan out over an increasing area as it moves away from the tiny channel to the target region where the ion will be absorbed as a whole on some small triggering site, or will not be absorbed at all on that site. The transit distance is estimated to be about 50 nanometers (Fogelson & Zucker 1985; Schweizer, Betz, & Augustine 1995), but the total distance traveled is increased many-fold by the diffusion mechanism. Thus the probability cloud becomes spread out over a region that is much larger than the size of the calcium ion itself, or of the trigger site. This spreading of the ion wave packet means that the ion may or may not be absorbed on the small triggering site.

Many different calcium ions contribute to the release of neurotransmitter from a vesicle. The estimated probability that a vesicle on a cerebral neuron will be released, per incident input action potential pulse, is far less than 100% (maybe only 50%). The very large quantum uncertainty at the individual calcium level ensures that this large empirical uncertainty of release entails that the quantum state of the nerve terminal will become a *quantum mixture* of states where the neurotransmitter is released, or, alternatively, is not released. This quantum splitting occurs at every one of the trillions of nerve terminals in the brain. This quantum splitting at each of the nerve terminals propagates, via the quantum mechanical process 2, first to neuronal behavior, and then to the behavior of the whole brain, so that, according to quantum theory, the state of the brain can become a cloudlike quantum mixture of many different classically describable brain states. In complex situations where the outcome at the classical level depends on noisy elements the corresponding quantum brain will evolve into a quantum mixture of the corresponding states.

The process 2 evolution of the brain is highly nonlinear, in the (classical) sense that small events can trigger much larger events, and that there are very important feedback loops. Some neurons can be on the verge of firing, so that small variations in the firing times of other neurons can influence whether or not this firing occurs. In a system with such a sensitive dependence on unstable elements, and on massive feedbacks, it is not reasonable to suppose, and not possible to demonstrate, that the process 2 dynamical evolution will lead generally to a single (nearly) classically describable quantum state. There might perhaps be particular special situations during which the massively parallel processing all conspires to cause the brain dynamics to become essentially deterministic and perhaps even nearly classically describable. But there is no likelihood that during periods of mental groping and uncertainty there cannot be bifurcation points in which one part of the quantum cloud of potentialities that represents the brain goes one way and the remainder goes another, leading to a quantum mixture of very different classically describable potentialities. The validity of the classical approximation certainly cannot be proved under these conditions, and, in view of the extreme nonlinearity of the neural dynamics, any claim that the large effects of the uncertainly principle at the synaptic level can never lead to quantum mixtures of macroscopically different states cannot be rationally justified.

What, then, is the effect of the replacement of a single, unique, classically described brain of classical physics by a quantum brain state composed of a mixture of several alternative possible classically describable brain states, each corresponding to a different possible experience?

A principal function of the brain is to receive clues from the environment, then to form an appropriate plan of action, and finally to direct the activities of the brain and body specified by the selected plan of action. The exact details of the chosen plan will, for a classical model, obviously depend upon the exact values of many noisy and uncontrolled variables. In cases close to a bifurcation point the dynamical effects of noise might, at the classical level, tip the balance between two very different responses to the given clues: e.g., tip the balance between the 'fight' or 'flight' response to some shadowy form, but in the quantum case one must allow and expect both possibilities at the macroscopic level a smear of classically alternative possibilities. The automatic mechanical process 2 evolution generates this smearing, and is in principle unable to resolve or remove it.

According to orthodox (von Neumann) quantum theory, achievement of a satisfactory reduction of the smeared out brain state to a brain state coordinated with the subject's streams of conscious experiences is achieved through the entry of a process 1 intervention, which selects from the smear of potentialities generated by the mechanical process 2 evolution a particular way of separating the physical state into a collection of components, each corresponding to some definite experience. The form of such an intervention is not determined by the quantum analog (process 2) of the physically deterministic continuous dynamical process of classical physics: some other kind of input is needed.

The choice involved in such an intervention *seems to us* to be influenced by consciously felt evaluations, and there is no *rational* reason why these conscious realities, which certainly *are* realities, cannot have the sort of effect that they seem to have.

5 Templates for Action

The feature of a brain state that tends to produce some specified experiential feedback can reasonably be expected to be a highly organized large-scale pattern of brain activity that, to be effective, must endure for a period of perhaps tens or hundreds of milliseconds. It must endure for an extended period in order to be able to bring into being the coordinated sequence of neuron firings needed to produce the intended feedback. Thus the neural (or brain) correlate of an intentional act should be something like a collection of the vibratory modes of a drumhead in which many particles move in a coordinated way for an extended period of time.

In quantum theory the enduring states are vibratory states. They are like the lowest-energy state of the simple harmonic oscillator discussed above, which tends to endure for a long time, or like the states obtained from such lowest-energy states by spatial displacements and shifts in velocity. Such states tend to endure as organized oscillating states, rather than quickly dissolving into chaotic disorder.

I call by the name 'template for action' a macroscopic brain state that will, if held in place for an extended period, tend to produce some particular action. Trial and error learning, extended over the evolutionary development of the species and over the life of the individual agent, should have the effect of bringing into the agent's repertoire of intentional process 1 actions the 'Yes–No' partitions such that the 'Yes' response will, if held in place for an extended period, tend to generate an associated recognizable feedback corresponding to the successful achievement of the intent. Successful living demands the generation through effort-based learning of templates for action.

My earlier discussion of the quantum indeterminacies that enter brain dynamics in association with the entry of calcium ions into the nerve terminals was given in order to justify the claim that the brain must be treated as a quantum system. However, the fact that quantum indeterminacies *enter* brain dynamics at the microscopic/ionic level does not mean that the process 1 interventions that are needed to link the evolving state of a person's brain to his or her conscious experiences must act microscopically. According to von Neumann's formulas, each process 1 intervention is specified by a set of nonlocal projection operators. This means that the effect of a process 1 action on a person's brain is generally *macroscopic*. Thus the quantum indeterminacies that enter brain dynamics at the microscopic/ionic level propagate via the Schroedinger equation (process 2) up to the macroscopic level where they produce a smear of potentialities that needs to be reduced to a form compatible with the occurrence of a conscious thought, if that thought is to enter a stream of consciousness. This dynamics expresses the core idea of the quantum theory of observation, which is that the reduction events are associated with increments in knowledge, and correspondingly reduce the physical state to the part of itself that is compatible with the knowledge entering a stream consciousness.

On the other hand, the only freedom provided by the quantum rules is the freedom to select the next process 1 action, and the instant at which it is applied. Thus a person's 'free choice' of what he or she *intends* to do can certainly enter the brain dynamics *at the macroscopic level*, but only as a process 1 action. This is where the 'latitude' offered by the quantum formalism, and associated with the 'free choice' of the experimenter emphasized by Bohr, enters the dynamics. This process 1 action can in fact be one whose 'Yes' alternative selects the set of brain states such that the template for the intended action is active. But this 'free choice' merely sets the stage for the entry of the statistical choice between the 'Yes' and 'No' alternatives whose relative statistical weights are specified by the quantum rules.

6 The Physical Effectiveness of Conscious Will and the Quantum Zeno Effect

A crucial question now arises: How does this dynamical psycho-neurological connection via process 1, *which can merely pose a question*, but not answer it, allow a person's effort to influence his or her physical actions?

Take an example. Suppose you are in a situation that calls for you to raise your arm. Associations via stored memories should elicit a brain activity having a component that when active on former occasions resulted in your experiencing your arm rise, and in which the template for arm-raising is active. According to the theory, this component of brain activity will, if sufficiently strong, cause an associated process 1 action to occur. This process 1 action will partition the quantum state of your brain in such a way that one component, labeled 'Yes', will be this component in which the arm-raising template is active. If the 'Yes' option is selected by nature then you will experience yourself causing your arm to rise, and the state of your brain will be such that the arm-raising template is active.

But the only dynamical freedom offered by the quantum formalism in this situation is the freedom to perform at a selected time some process 1 action. Whether or not the 'Yes' component is actualized is determined by 'nature' on the basis of a statistical law. So the effectiveness of the 'free choice' of this process 1 in achieving the desired end would generally be quite limited. The net effect of this 'free choice' would tend to be nullified by the randomness in nature's choice between 'Yes' and its negation 'No'.

A well-known non-classical feature of quantum theory provides, however, a way to overcome this problem, and convert the available 'free choices' into effective mental causation.

6.1 The Quantum Zeno Effect

A well studied feature of the dynamical rules of quantum theory is this: Suppose a process 1 query that leads to a 'Yes' outcome is followed by a rapid sequence of very similar process 1 queries. That is, suppose a sequence of identical or very similar process 1 actions is performed, that the first outcome is 'Yes', and that the actions in this sequence occur in very rapid succession on the time scale of the evolution of the original 'Yes' state. Then the dynamical rules of quantum theory entail that the sequence of outcomes will, with high probability, all be 'Yes': the original 'Yes' state will, with high probability, be held approximately in place by the rapid succession of process 1 actions, even in the face of very strong physical forces that would, in the absence of this rapid sequence of actions, quickly cause the state to evolve into some very different state (Stapp 2004a, Sect. 12.7.3).

The *timings* of the process 1 actions are, within the orthodox formulations, controlled by the 'free choices' on the part of the agent. Mental effort applied to a conscious intent increases the intensity of the experience. Thus it is consistent and reasonable to suppose that the rapidity of a succession of essentially identical process 1 actions can be increased by mental effort. But then we obtain, as a mathematical consequence of the basic dynamical laws of quantum mechanics described by von Neumann, a potentially powerful effect of mental effort on the brain of the agent! Applying mental effort increases the rapidity of the sequence of essentially identical intentional acts, which then causes the template for action to be held in place, which then produces the brain activity that tends to produce the intended feedback.

This 'holding-in-place' effect is called the quantum Zeno effect, an appellation that was picked by the physicists E.C.G. Sudarshan and R. Misra (1977) to highlight a similarity of this effect to the 'arrow' paradox discussed by the fifth century B.C. Greek philosopher, Zeno the Eleatic. Another name for this effect is 'the watched-pot effect'.

The quantum Zeno effect can, in principle, hold an intention and its template in place in the face of strong mechanical forces that would tend to disturb it. This means that agents whose mental efforts can sufficiently increase the rapidity of process 1 actions would enjoy a survival advantage over competitors that lack such features. They could sustain beneficial templates for action in place longer than competitors who lack this capacity. Thus the dynamical rules of quantum mechanics *allow* conscious effort to be endowed with the causal efficacy needed to permit its deployment and evolution via natural selection.

6.2 William James's Theory of Volition

This theory was already in place when a colleague, Dr. Jeffrey Schwartz, brought to my attention some passages from *Psychology: The Briefer Course*, written by William James. In the final section of the chapter on Attention, James (1892) writes:

> I have spoken as if our attention were wholly determined by neural conditions. I believe that the array of things we can attend to is so determined. No object can catch our attention except by the neural machinery. But the amount of the attention which an object receives after it has caught our attention is another question. It often takes effort to keep mind upon it. We feel that we can make more or less of the effort as we choose. If this feeling be not deceptive, if our effort be a spiritual force, and an indeterminate one, then of course it contributes coequally with the cerebral conditions to the result. Though it introduce no new idea, it will deepen and prolong the stay in consciousness of innumerable ideas which else would fade more quickly away. The delay thus gained might not be more than a second in duration – but that second may be critical; for in the rising and falling considerations in the mind, where two associated systems of them are nearly in equilibrium it is often a matter of but a second more or less of attention at the outset, whether one system shall gain force to occupy the field and develop itself and exclude the other, or be excluded itself by the other. When developed it may make us act, and that act may seal our doom. When we come to the chapter on the Will we shall see that the whole drama of the voluntary life hinges on the attention, slightly more or slightly less, which rival motor ideas may receive.

In the chapter on Will, in the section entitled *Volitional Effort is Effort of Attention*, James writes:

> Thus we find that we reach the heart of our inquiry into volition when we ask by what process it is that the thought of any given action comes to prevail stably in the mind.

And later

> The essential achievement of the will, in short, when it is most 'voluntary', is to attend to a difficult object and hold it fast

> before the mind. [...] Effort of attention is thus the essential phenomenon of will.

Still later, James says:

> Consent to the idea's undivided presence, this is effort's sole achievement. [...] Everywhere, then, the function of effort is the same: to keep affirming and adopting the thought which, if left to itself, would slip away.

James apparently recognized the incompatibility of these pronouncements with the physics of his day. At the end of *Psychology: The Briefer Course*, he said, presciently, of the scientists who would one day illuminate the mind–body problem:

> The best way in which we can facilitate their advent is to understand how great is the darkness in which we grope, and never forget that the natural-science assumptions with which we started are provisional and revisable things.

It is a testimony to the power of the grip of old ideas on the minds of scientists and philosophers alike that what was apparently evident to William James already in 1892 – namely that a revision of the mechanical precepts of nineteenth century physics would be needed to accommodate the structural features of our conscious experiences – still fails to be recognized by many of the affected professionals even today, more than three-quarters of a century after the downfall of classical physics, apparently foreseen by James, has come, much-heralded, to pass.

James's description of the effect of volition on the course of mind–brain process is remarkably in line with what had been proposed, independently, from purely theoretical considerations of the quantum physics of this process. The connections described by James are explained on the basis of the same dynamical principles that had been introduced by physicists to explain atomic phenomena. Thus the whole range of science, from atomic physics to mind–brain dynamics, is brought together in a single rationally coherent theory of a world that is constituted not of matter, as classically conceived, but rather of an informational structure that causally links the two elements that combine to constitute actual scientific practice, namely the psychologically described contents of our streams of conscious experiences and the mathematically described objective tendencies that tie our chosen actions to experience.

No comparable success has been achieved within the framework of classical physics, in spite of intense efforts spanning more than three centuries. The reasons for this failure are easy to see: classical physics systematically exorcizes all traces of mind from its precepts, thereby banishing any logical foothold for recovering mind. Moreover, according to quantum physics all causal effects of consciousness act within the latitude provided by the uncertainty principle, and this latitude shrinks to zero in the classical approximation, eliminating the causal effects of consciousness.

7 Support from Contemporary Psychology

A great deal has happened in psychology since the time of William James. However, many psychologists, neuroscientists, and philosophers who intended to stay in tune with the basic precepts of physics became locked to the ideas of nineteenth century physicists and failed to acknowledge or recognize the jettisoning by twentieth century physicists of classical materialism and the principle of the causal closure of the physical. Thus while the physicists were bringing effects attributed to the conscious intentions of human agents into the dynamical description of the physically described world, mainline psychologists, embracing behaviorism, sought to remove such features even from psychology, and most philosophers of mind followed suit.

The eventual failure of the behaviorist program to account for the facts of human behavior, and in particular for linguistic behavior, led to the rehabilitation of 'attention' during the 1950s, and many hundreds of experiments have been performed during the past fifty years for the purpose of investigating empirically those aspects of human behavior that we ordinarily link to our consciousness. So we can now inquire: How well does the above-described quantum-theory-based approach to mind–brain dynamics square with these newer data?

Harold Pashler's 1998 book *The Psychology of Attention* describes a great deal of this empirical work, as well as the intertwined theoretical efforts to understand the nature of an information-processing system that could account for the fine details of the empirical data. Two key concepts are the notions of 'attention' and of a processing 'capacity'. The former is associated with an internally directed selection between different possible allocations of the available processing 'capacity'. A third concept is 'effort', which is empirically linked to incentives, and to reports by subjects of 'trying harder'. Effort increases the portion of the processing capacity that is being applied to a cognitively directed task.

Pashler organizes his discussion by separating perceptual processing from post-perceptual processing. The former covers processing that, first of all, identifies such basic physical properties of stimuli as location, color, loudness, and pitch, and, secondly, identifies stimuli in terms of categories of meaning. The post-perceptual process covers the tasks of producing motor actions and cognitive action beyond mere categorical identification. Pashler emphasizes (p. 33) that "the empirical findings of attention studies specifically argue for a distinction between perceptual limitations and more central limitations involved in thought and the planning of action". The existence of these two different processes, with different characteristics, is a principal theme of Pashler's book (pp. 33, 263, 293, 317, 404)

Orthodox quantum theory also features two separate processes. Quantum theory, applied to the mind–brain system, in accordance with von Neumann's formulation, involves, first, the unconscious mechanical brain process called process 2. A huge industry has developed that traces these essentially classically describable processes in the brain. But, according to orthodox contemporary physics, another process, von Neumann's process 1, must also enter into the causal structure. Its physical effects can become manifest in connection with an impulsive feeling described as 'effort'. The effect of this 'effort of attention' is to inject into brain activity, and thence eventually into overt behavior, effects of *intentional inputs*.

Two kinds of process 1 actions are possible. One kind would be determined by brain activity alone. It would be the kind of action associated with James's assertion that "no object can catch our attention except by the neural machinery". However, another kind of process 1 action is possible within the framework provided by von Neumann's formulation. It can stem from a *positive evaluation* based on the felt or experiential quality of internal coherence, and would tend to make the process 1 psychophysical event in which it occurs immediately repeat itself a short time later, with the rapidity of these repeated actions being increasable, *up to a certain limit*, by an experienced quality of the event called 'effort'. Such a process 1 action could, within the orthodox quantum framework, induce a rapid sequence of similar actions that could activate a quantum Zeno effect that would effectively inject a rapid sequence of mental intentions into the course of brain activity.

This quantum conceptualization of the action of mind on brain is, as we shall now see, in good accord with the *details* of the data described by Pashler. Those data did not necessarily – *from non-quantum considerations* – need to have the detailed structure that it is empirically found to have. Indeed, the various classical-type theories examined by Pashler did not entail it. Consequently, these data provides some empirical support for this quantum-physics-based idea of the mind–brain connection

The 'perceptual' aspect of brain process discussed by Pashler can be associated with process 2, and also with the essentially passive-type process 1, whereas the higher-level processing that Pashler identifies can be associated with the active mode of process 1.

The *perceptual* aspects of the data described by Pashler can, I believe, be accounted for by essentially classical parallel mechanical processing. But it is the high-level processing, which is linked to active mental effort, that is of prime interest here. The data pertaining to this second kind of process are the focus of Part II of Pashler's book.

Examination of Part II of Pashler's book shows that the quantum-physics-based theory accommodates naturally all of the detailed structural features of the empirical data that he describes. He emphasizes (p. 33) a specific finding, namely strong empirical evidence for what he calls a *central processing bottleneck* associated with the attentive selection of a motor action. This kind of bottleneck is what the quantum-physics-based theory predicts: the bottleneck is precisely the single linear sequence of process 1 actions that enters so importantly into the quantum theoretic description of the mind–matter connection.

The sort of effect that Pashler finds is illustrated by a result he describes that dates from the nineteenth century: mental exertion reduces the amount of physical force that a person can apply. He notes that "this puzzling phenomena remains unexplained" (p. 387). However, it is a natural consequence of the physics-based theory: creating physical force by muscle contraction requires an effort that opposes the natural dissipative physical tendencies generated by process 2. This opposing tendency is produced by the quantum Zeno effect, and should be roughly proportional to the number of bits per second of central processing capacity that is devoted to the task. So if part of this processing capacity is directed to another task, then the muscular force will diminish.

An interesting experiment mentioned by Pashler involves the simultaneous tasks of doing an IQ test and giving a foot response to rapidly presented sequences of tones of either 2000 or 250 Hz. The subject's mental age, as measured by the IQ test, was reduced from adult to 8 years. Effort can be divided, but at a maximal level there is a net total rate of effortful process 1 action.

Another interesting experiment showed that, when performing at maximum speed, with fixed accuracy, subjects produced responses at the same rate whether performing one task or two simultaneously:

the limited capacity to produce responses can be divided between two simultaneously performed tasks (p. 301).

Pashler also notes (p. 348) that "recent results strengthen the case for central interference even further, concluding that *memory retrieval* is subject to the same discrete processing bottleneck that prevents simultaneous response selection in two speeded choice tasks".

In the section on *Mental Effort*, Pashler reports that "incentives to perform especially well lead subjects to improve both speed and accuracy", and that the motivation had "greater effects on the more cognitively complex activity". This is what would be expected if incentives lead to effort that produces increased rapidity of the events, each of which injects mental intent into the physical process.

Studies of sleep-deprived subjects suggest that in these cases "effort works to counteract low arousal". If arousal is essentially the rate of occurrence of conscious events then this result is what the quantum model would predict.

Pashler notes that "performing two tasks at the same time, for example, almost invariably [...] produces poorer performance in a task and increases ratings in effortfulness". And "increasing the rate at which events occur in experimenter-paced tasks often increases effort ratings without affecting performance". "Increasing incentives often raises workload ratings and performance at the same time." All of these empirical connections are in line with the general principle that effort increases the rate of conscious events, each of which inputs a mental intention, and that this resource can be divided between tasks.

After analyzing various possible mechanisms that could cause the central bottleneck, Pashler (pp. 307–8) says "the question of why this should be the case is quite puzzling". The citing of these data is meant only to indicate that these data *are in natural concordance* with the structure of orthodox (von Neumann) quantum mechanics, supplemented by the idea that mental effort can, *by virtue of the known quantum laws*, tend to hold attention in place, and thus tend to instigate consciously intended physical actions. Citing these data is *not* intended to demonstrate that von Neumann quantum mechanics is the *only* possible way to explain these empirical findings. Still, orthodox von Neumann quantum theory does provide the foundation for a natural physics-based causal explanation of these complex data that is in line with our normal intuition that our conscious efforts can influence our physical actions. Adopting orthodox von Neumann quantum theory allows one to avoid the gross philosophical contortions that have been proposed in order to reconcile the apparent physical efficacy of conscious effort with classical/materialist theories that entail the causal closure of the physically described aspects.

8 Application to Neuropsychology

The most direct evidence pertaining to the effects of conscious choices upon brain activities comes from experiments in which consciously controlled cognitive efforts are found to be empirically correlated to measured physical effects in the brain. An example is the experiment of Ochsner et al. (2001). The subjects are trained how to cognitively re-evaluate emotional scenes by consciously creating and holding in place an alternative fictional story of what is really happening in connection with an emotion-generating scene they are viewing.

> The trial began with a 4-second presentation of a negative or neutral photo, during which participants were instructed simply to view the stimulus on the screen. This interval was intended to provide time for participants to apprehend complex scenes and allow an emotional response to be generated that participants would then be asked to regulate. The word 'attend' (for negative or neutral photos) or 'reappraise' (negative photos only) then appeared beneath the photo and the participants followed this instruction for 4 seconds.
>
> To verify whether the participants had, in fact, reappraised in this manner, during the post-scan rating session participants were asked to indicate for each photo whether they had reinterpreted the photo (as instructed) or had used some other type of reappraisal strategy. Compliance was high: On less than 4% of trials with highly negative photos did participants report using another type of strategy.

Reports such as these can be taken as evidence that the streams of consciousness of the participants do exist and contain elements identifiable as efforts to reappraise.

Patterns of brain activity accompanying reappraisal efforts were assessed by using functional magnetic imaging resonance (fMRI). The fMRI results were that reappraisal was positively correlated with increased activity in the left lateral prefrontal cortex and the dorsal medial prefrontal cortex (regions thought to be connected to cognitive control) and decreased activity in the (emotion-related) amygdala and medial orbito-frontal cortex.

How can we explain the correlation revealed in this experiment between the mental reality of 'conscious effort' and the physical reality of measured brain behavior?

According to the precepts of classical physics, the subject's behavior is controlled by physically described variables alone, and his feeling that his 'conscious effort' is affecting his thinking is an illusion: the causal chain of physical events originating in the instructions being fed to the trained subject is controlling the brain response, and his feeling of 'conscious effort' is an epiphenomenal side-effect that has no effect whatever on his brain.

The validity of that picture cannot be empirically verified or confirmed: it is an unverifiable conjecture. Nor has this conjecture any rational foundation in science or basic physics. The conjecture originates from the classical principle of the causal closure of the physical, which does not generally hold in quantum theory. That principle rests on a classical-physics-based bottom-up determinism that starts at the elementary particle level and works up to the macro-level. But, according to the quantum principles, the determinism at the bottom (ionic) level fails badly in the brain. The presumption that it gets restored at the macro-level is wishful and unprovable.

According to quantum mechanics, the microscopic uncertainties must rationally be expected to produce, via the Schroedinger equation (of brain plus environment), macroscopic variations that, to match observation, need to be cut back by quantum reductions. This means process 1 interventions. This leads, consistently and reasonably, to the entry of mental causation as described above, where the subject's conscious effort is *actually* causing what his conscious understanding *believes*, on the basis of life-long experience, that effort to be causing.

There is no rational explanation for the existence of the 'illusion of conscious influence' when no such influence exists, but a completely reasonable explanation for the subject's believing that his conscious effort has an influence when that experienced effort has an influence that incessantly demonstrates itself to the subject.

As regards causation, the structure of quantum theory effects a *replacement*, within the dynamics, of what is *unknowable in principle*, namely the empirically inaccessible microscopic features of the brain, by data of a different kind, which *are* knowable in principle, namely our efforts. This replacement of inaccessible-in-principle data by accessible-in-practice data leads to statistical predictions connecting empirically describable conscious intentions to empirically describable perceptual feedbacks. The psychologically described and mathematically described components of the theory become cemented together by quantum rules that work in practice.

What is the rational motivation for adhering to the classical approximation?

The applicability of the classical approximation to this phenomenon certainly does not follow from physics considerations: calculations based on the known properties of nerve terminals indicate that quantum theory must in principle be used. Nor does it follow from the fact that classical physics works reasonably well in neuroanatomy and neurophysiology: quantum theory explains why the classical approximation works well in those domains. Nor does it follow rationally from the massive analyses and conflicting arguments put forth by philosophers of mind. In view of the turmoil that has engulfed philosophy during the three centuries since Newton's successors cut the bond between mind and matter, the re-bonding achieved by physicists during the first half of the twentieth century must be seen as a momentous development. Ignoring in the scientific study of the mind–brain connection this enormously pertinent development in basic science appears to be, from a scientific perspective, an irrational choice.

The materialist claim is that *someday* the mind will be understood to be the product of completely mindless matter. Karl Popper called this prophecy "promissory materialism". But can these connections reasonably be expected to be understood in terms of a physical theory that is known to be false, and, moreover, to be false because it is an approximation that eliminates a key feature of the object of study, namely the causal effects of mental effort upon brain activity.

The only objections I know to applying the basic principles of orthodox contemporary physics to brain dynamics are, first, the forcefully expressed opinions of some non-physicists that the classical approximation provides an entirely adequate foundation for understanding mind–brain dynamics, in spite of quantum calculations that indicate just the opposite; and, second, the opinions of some conservative physicists, who, apparently for philosophical reasons, contend that the practically successful orthodox quantum theory, which is intrinsically dualistic, should, be replaced by a theory that re-converts human consciousness into a causally inert witness to the mindless dance of atoms, as it was in 1900. Neither of these opinions has any rational basis in contemporary science, as will be further elaborated upon in the sections that follow. And they leave unanswered the hard question: Why should causally inert consciousness exist at all, and massively deceive us about its nature and function?

9 Roger Penrose's Theory and Quantum Decoherence

Increased interest in quantum mechanical theories of mind has been kindled by two recent books by Roger Penrose. These books, *The Emperor's New Mind*, and *Shadows of the Mind*, along with a paper by Hameroff and Penrose (1996), propose a quantum theory of consciousness that, like the present one, is based on von Neumann's formulation of quantum theory. But the Penrose–Hameroff theory brings in some controversial ideas that are not used in the more direct application of orthodox quantum mechanics described in this book.

An essential difference between the present proposal and that of Penrose and Hameroff is that their theory depends on the assumption that a property called 'quantum coherence' extends over a large portion of the brain, whereas the theory described here does not. This property is a technical matter that I do not want to enter into right here, beyond remarking that most quantum physicists deem it highly unlikely that the quantum coherence required by the Penrose–Hameroff theory could be sustained in a warm, wet, living brain. Quantitative estimates that appear to back up this negative opinion have been made by Tegmark (2000). A rebuttal has been offered by Hagen, Hameroff, and Tuszynski (2002), but the needed level of coherence still looks very difficult to achieve.

The expected (by most physicists) lack of long-range quantum coherence in a living brain is, in fact, a great asset to the von Neumann approach described in this book. This lack of coherence (decoherence) means that the quantum brain can be conceived to be, to a very good approximation, simply a collection of classically conceived alternative possible states of the brain. The point here is that the interaction with the environment effectively washes out all observable effects of the possible-in-principle interferences between parts of the brain that are spatially separated by an appreciable distance: the only quantum effects that survive decoherence are those associated with very close neighbors. Thus the quantum state of the brain is effectively, to a very good approximation, simply a collection of alternative possible classically described brains. They all exist together as 'parallel' parts of a potentiality for future additions to a stream of consciousness. The residual quantum effects arise from the fact that these quasi-classical 'parallel' brain states are allowed to interact with their very close neighbors. Still, these surviving linkages to close neighbors make the quantum model significantly different in principle from a purely classical model: no classical possibility can interact with an *alternative* classical possibility, no matter how close together they are.

The only macroscopic quantum effect that appears to survive the decoherence effects is the quantum Zeno effect. This permits neuroscientists unfamiliar with quantum theory to have a very accurate, simple, intuitive idea of the quantum state of a brain. It can be imagined to be an evolving set of nearly classical brains with, however, the following four non-classical properties:

1. Each almost-classical possibility is slightly smeared out in space relative to a strictly classical idealization, and it fans out in accordance with the uncertainty principle.
2. At each occurrence of a conscious thought, the set of possibilities is reduced to the subset compatible with the occurring increment of knowledge.
3. Microscopic chemical interactions are treated quantum mechanically.
4. In the presence of effortful intent, the quantum Zeno effect acts to keep the associated template for action in place for longer than classical mechanics would allow.

A second principal difference between the Penrose–Hameroff theory and the one being described here is that the former depends on the complex question of the nature of quantum gravity, which is currently not under good theoretical control, whereas the present approach is based only on the *fundamental principles of orthodox quantum theory*, which, thanks to the efforts of John von Neumann, are under good control. Penrose's proposal strongly links consciousness to the *gravitational* interactions of parts of the brain with other parts of the same brain, whereas the theory being advanced here supposes gravitational interactions between parts of the same brain to be negligible.

The third difference is that Penrose's approach involves a very much disputed argument that claims to *deduce* from (1) the fact that mathematicians construct proofs that they believe to be valid, and from (2) some deep mathematical results due to Kurt Gödel, the conclusion that brain processes *must* involve a non-algorithmic (not discretely describable) process. According to the present approach, contemporary orthodox quantum theory already requires the physically described process 2 aspects of brain processes to be influenced by process 1 interventions coming from streams of consciousness. The theory leaves open the important question of how these interventions, which are *treated* pragmatically simply as experimenter-selected choices of boundary conditions, come to be what they turn out to be: this is the causal gap! These interventions are not required by present understanding to be governed by algorithmic processes.

10 Non-Orthodox Versions of Quantum Theory and the Need for Process 1

Eugene Wigner introduced the term 'orthodox' to describe von Neumann's formulation of quantum theory. I use the term more broadly to include, *at the pragmatic level*, also the Copenhagen formulation. But at the *ontological* level I mean the von Neumann–Tomonaga–Schwinger description that includes the entire physical universe in the physically described quantum world, and that accepts the occurrence of the process 1 interventions in the process 2 evolution of the physically described state of the universe.

This conventional formulation of quantum theory – with experimenter-induced interventions – is the one used in practice by experimental physicists who need to compare the predictions of the theory to empirical data. It is consequently the form of the theory that is actually supported by the empirical facts.

It might seem odd, therefore, that any quantum physicist would want to promote an alternative formulation. It seems particularly strange that there could be physicists who now seek to remove the effects of consciousness from the basic dynamics; physicists who want to reverse the great twentieth century achievement of rescuing consciousness from the passive limbo to which it had been consigned by classical physical principles. It seems strange that there could be physicists who seek to retreat from the idea of giving consciousness a causal role that: 1) accords with our deep intuitions; 2) meshes neatly with empirical practice; 3) explains naturally the effortful learning of new tasks; and that 4) allows consciousness to evolve by natural selection, by virtue of its capacity to aid our bodily survival. Yet theoretical physicists who favor such a reversion do in fact exist.

The feature of quantum theory that precipitates the disagreements among physicists is that it is exceedingly difficult to detect directly by physical measurements whether a large physical system that is strongly interacting with its environment is, or is not, acting as a quantum agent: it is virtually impossible to determine, directly by measurements, whether reduction events are occurring in such a system. Given such empirical latitude it is natural that theorists should tend to build alternative theories. And from the perspective of a theoretical physicist it is of course desirable to have a causal structure that is completely fixed in terms of the purely physical descriptions with which he or she is familiar, in spite of the deep problems that such a restriction eventually generates, both mathematically and philosophically.

There are three main non-orthodox approaches to the problem of imbedding pragmatically validated quantum theory in some coherent conception of reality itself. These are the many-worlds approach initiated by Everett (1957), the pilot-wave approach of Bohm (1952, 1993) and the spontaneous-reductions approach of Ghirardi, Rimini, and Weber (1986).

The many-worlds approach is the most radical and sweeping. It asserts that the quantum state of the physical universe exists and evolves *always* under the exclusive control of the local deterministic process 2. In this scheme no reduction events occur at the level of objective reality itself. The fact that we *seem* to choose particular experiments that *seem* to have outcomes that conform to the predictions of quantum theory then needs to be explained as essentially some kind of persisting subjective illusion that produces coherent long-term streams of human *conscious* events that somehow conform over long times to the statistical predictions of the orthodox theory, even though the *physical reduction events that logically entail these properties in the orthodox approach are now asserted not to occur*. The *consciously perceived* experiences that conform to the statistical rules of pragmatic quantum theory then need to be explained as intricate properties of the purely mental by-products of a continuous physical process that eschews the interventions and reductions that provide the mathematical foundation of the orthodox understanding of the empirical facts.

The pilot-wave approach claims that there really is *both* a world of the kind specified in classical physics – a world that determines the content of our human streams of conscious experience – and *also* a real state of the universe of the kind specified in quantum theory. It asserts that this latter world always evolves via process 2, with no collapses or interventions, and that the real classical world is buffeted around by the real quantum world in a way that accounts for the validity of the empirical predictions of pragmatic quantum theory.

The spontaneous-reductions approach maintains that the evolution via the local mechanical process 2 is interrupted from time to time by a sudden spontaneous and random reduction event that keeps the physical universe, at the visible level, roughly in accord with the precepts of classical physics, while allowing quantum processes to work, virtually undisturbed, at the microscopic level of atomic physics.

All three of these approaches differ fundamentally from the orthodox Heisenberg–von Neumann approach described above in that they adhere to the principle of the causal closure of the physical, apart perhaps from the entry of some purely random elements. In particular, they exclude any causal effects of our conscious minds. However, all three run into serious technical difficulties.

10.1 The Many-Worlds (or Many-Minds) Approach and Decoherence

I received recently a query from a colleague, who wrote:

> I would appreciate your answering a question I have. There is much disagreement in the literature about the reduction process and how it works, including controversy over whether there is any such thing as reduction. I have read numerous statements from physicists that measurement involves interaction of a quantum system with its environment, and is (it is asserted) therefore 'nothing but' Schroedinger [process 2] evolution on a larger system.

It is indeed sometimes claimed that the interaction of a system with its environment effectively solves the 'measurement' problem (which is essentially the problem of how to connect the physically/mathematically described aspects of quantum theory to human experience). However, the principal investigators of the effects of this interaction (e.g., E. Joos, 1996; D. Zeh, 1996; W. Zurek, 2002) make no such strong claim. Joos (p. 3) emphasizes that even when the interaction with the environment is included one is left with not one single classical world but with a host of possible classical worlds "thus leaving the measurement problem essentially unsolved (unless one is willing to accept some variant of the Everett interpretation)". Zeh (p. 17), commenting on the problems that remain after the interaction with the environment has been included, says: "A way out of this dilemma in terms of the wave function itself seems to require one of the following two possibilities: (1) a modification of the Schroedinger equation that explicitly describes a collapse, or (2) an Everett type of solution, in which all measurement outcomes are assumed to coexist in one formal superposition, but to be perceived separately as a consequence of their dynamical decoupling." This 'Everett type of solution' is usually called a many-worlds or a many-minds solution.

Zurek (p. 5) says:

> At first glance, the Many Worlds and Copenhagen Interpretations have little in common. The Copenhagen Interpretation demands an a priori 'classical domain' with a border that enforces a classical 'embargo' by letting through just one potential outcome. The Many Worlds Interpretation aims to abolish the need for a border altogether. Every potential outcome is accommodated in the ever-proliferating branches of the wave function of the Universe. The similarity between the difficulties faced by these two viewpoints becomes apparent, nevertheless, when we ask the obvious question, "Why do I, the observer, perceive only one of the outcomes?" Quantum theory with its freedom to rotate bases in the Hilbert space, does not even define which states of the universe correspond to the 'branches'. Yet our perception of a reality with alternatives – not a coherent superposition of alternatives – demands an explanation of when, where, and how it is decided what the observer actually records. Considered in this context, the Many Worlds Interpretation in its original version does not really abolish the border but pushes it all the way to the boundary between the physical universe and consciousness. Needless to say, this is a very uncomfortable place to do physics.

Later on (pp. 20–21) he returns to this problem: "Why do we perceive just one of the quantum alternatives?"; "the process of decoherence we have described above is bound to affect the states of the brain [...] decoherence applies to our own 'state of mind'"; "There is little doubt that the process of decoherence sketched in this paper is an important element of the big picture [...]. There is even less doubt that this rough outline will be further extended. Much work needs to be done, both on technical issues [...] and on problems that require new conceptual input (such as [...] answering the question of how an observer fits into the big picture)."

These comments make clear the fact that interaction with the environment (and the resulting technical effect known as environmental decoherence) does not by itself solve the measurement problem, namely the problem of accounting for the fact that an observer perceives just one classically describable world, not the continuous collection of them generated by process 2 acting alone – *which includes all effects of the environment*.

The question, then, is whether the many-worlds/minds option is rationally acceptable. I have described (in Stapp 2002) a specific difficulty with the many-worlds approach that is sufficiently serious to block, at the present time, the claim that the Schroedinger equation alone (i.e., process 2), including all interactions with the environment, is sufficient – without process 1, or some surrogate of process 1 – to tie the quantum mathematics to testable predictions about human experiences. Such predictions are required for the theory to be scientifically meaningful, and they are obtained in the Copenhagen/von Neumann orthodox approach only by bringing in process 1 interventions.

The reason, in brief, why process 1, or something that does the same job, seems to be needed is this: If the universe has been evolving since the big bang solely under the influence of the Schroedinger equation – i.e., process 2 – then every object and every human brain would by now, due to the uncertainty conditions on the original positions and velocities, be represented in quantum theory by an amorphous continuum; the center point of each object would not lie at a particular point, or even be confined to a small region, but would be continuously spread out over a huge region. Likewise, the state of the brain of every observer of this object would be a smeared out conglomeration of many different classical-type brains. That is, if a human person were observing an object, whose center point, as specified by its quantum state, were spread out over a region several meters in diameter, then the state of the brain of that person would have, for each of these different locations, a part corresponding to the observer's seeing the object in that location. If each of these parts of the brain were accompanied by the corresponding experience, then there would exist not just one experience corresponding to seeing the object in just one place, but a continuous aggregation of experiences, with one experience for each of the possible locations of the object in the large region. Thus this theory is often called, quite rightly, a 'many-minds' interpretation: each person's brain evolves quickly into – and in fact would never be other than – a smeared out continuum, and each stream of consciousness would be part of a continuous blur of classically describable possibilities.

In order to extract from quantum theory a set of predictions pertaining to human experiences, and hence to give empirical meaning to the theory, this smeared out collection of different brain structures must be resolved in a very special way into a collection of discrete parts, each corresponding to one possible experience. This discreteness condition is a technical point, but it constitutes the essential core of the measurement problem. Hence I must explain it! It is often called the measurement problem, and is the problem of relating the quantum mathematics to the empirically observed phenomena.

Evolution according to the Schroedinger equation (process 2) generates in general, as I have just explained, a state of the brain of an observer that is a smeared out continuum of component parts. One cannot assign a nonzero probability to each one of such a continuum of possibilities, because the total probability would then be infinity, instead of one (unity). However, the mathematical rules of quantum theory have a well-defined way to deal with this situation: they demand that the space of possibilities be divided in a certain very restrictive way into a countable set of alternative possibilities, where a 'countable' set is a set that can be numbered (i.e., placed in one-to-one correspondence with the whole numbers $1, 2, 3, \ldots$, or with some finite subset of these numbers). The need to specify a particular countable set of parts is the essential problem in the construction of a satisfactory quantum theory. It is the problem cited by Zurek when he said: "Quantum theory with its freedom to rotate bases in the Hilbert space, does not even define which states of the universe correspond to the 'branches'." But then the technical problem that the many-worlders must resolve is this: How does one specify a satisfactory particular countable set of different brain states from process 2 alone, when process 2 is a continuous local process that generates a structure that continuously connects components that correspond to very different experiences, and hence must belong to different members of the countable set? The problem is to divide a continuum of brain states into a countable set of discrete (and orthogonal) components by means of the strictly continuous process 2 alone, and in a way such that the distinct parts correspond to distinguishable experiences.

Copenhagen quantum theory accomplishes this selection of a preferred set of discrete states by means of an intervention by the experimenter. In the simplest case the countable set of distinguishable experiences has just two elements, 'Yes' and 'No'. The experimenter selects a particular probing action that picks out from the continuously infinite set of possible queries some particular one. In this way, the basic problem of specifying a countable set of discrete parts is solved by bringing into the theory definite choices on the part of the experimenter. Von Neumann solves this discreteness problem in this same way, and gives the physical manifestation of this crucial agent-dependent selection process the name 'process 1'.

Einstein (1951, p. 670) posed essentially the same problem in a clear way. Suppose a pen that draws a line on a moving scroll is caused to draw a blip when a radioactive decay is detected by some detector. If the only process in nature is process 2, then the state of the scroll will be a blurred out state in which the blip occurs in a continuum of alternative possible locations. Correspondingly, the brain of a person who is observing the scroll will be in a smeared out state containing a continuously connected collection of components, with one component corresponding to each of the possible locations of the blip on the scroll. But how does this smeared out continuously connected state of the brain get divided by process 2 alone into components to which well-defined probabilities can be assigned? The quantum statistical predictions cover only those cases in which there is a specified countable collection of distinct possibilities.

A key feature of the orthodox approach is the 'empirical fact' that experimenters do have definite thoughts, and that they can therefore choose to place the devices in definite locations. Thus it is the empirically experienced discreteness of the choice made by the experimenter that resolves the discreteness problem. But an experimenter represented exclusively by a state governed solely by process 2 has nothing discrete about him: his brain is a continuous smear with no dynamically defined dividing lines.

The founders of quantum theory (and von Neumann) recognized this basic problem of principle, and in order to resolve it went to a radical and revolutionary extreme: they introduced human experimenters with efficacious free choices into the physical theory. This was a giant break from tradition. But the enormity of the problem demanded drastic measures. Because such powerful thinkers as Wolfgang Pauli and John von Neumann found it necessary to embrace this revolutionary idea, anyone who claims that this unprecedented step was wholly unnecessary certainly needs to carefully explain why. This has not yet been done. (See the next chapter for further elaboration.)

Although bringing the consciousness of human agents into the dynamics is certainly quite contrary to the ideas of classical physics, the notion that our streams of consciousness play a causal role in the determination of our behavior is not outlandish: it is what one naturally expects on the basis of everyday experience. Orthodox quantum theory solves a serious *technical* problem in a way that automatically allows, as a by-product, our conscious thoughts to causally affect our physical actions in the way that they seem to us to do!

10.2 Bohm's Pilot-Wave Model

Bohm's pilot-wave model (Bohm 1952) is an attempt to supplement process 2 by adding an extra *mechanical* element, not involving mind, that does the job that the mind-driven process 1 does in the orthodox interpretation.

Bohm's model is built on a resuscitation of the classical idea of a world of point particles (atomic-sized planet-like objects) and classical fields, such as the electric and magnetic fields. The function of his postulated world of classically conceived particles and fields is to determine, in accordance with classical concepts, what our experiences will be. Because there is, according to Bohm, only one such classical world, there will be, by fiat, only one experience, not the infinite continuum of them that the process-2-controlled wave function by itself would seem to generate.

I once asked Bohm how he answered Einstein's charge that his model was 'too cheap'. He said that he completely agreed! Notice, in this connection, that in the last two chapters of his book with Hiley, Bohm goes beyond this simple model, and tries to come to grips with the deeper problems that are being considered here by introducing the notions of implicate and explicate order, But those extra ideas are considerably less mathematical, and much more speculative and vague, than the pilot-wave model that many other physicists want to take more seriously than did Bohm himself.

Bohm certainly appreciated the need to deal more substantively with the problem of consciousness. He wrote a paper on the subject (Bohm 1986, 1990), which ended up associating consciousness with an infinite tower of pilot waves, each one piloting the wave below. But the great virtue of the original pilot-wave model, namely the fact that it was simple and deterministic with cleanly specified solvable equations, became lost in this infinite tower.

Over and beyond these problems with consciousness there is a serious technical problem: a Bohm-type deterministic model apparently cannot be made to accommodate relativistic particle creation and annihilation, which is an important feature of the actual world in which we live. Completing the ontology by adding a classically conceived mechanistically determined world – instead of choices made by agents and by nature – has never been satisfactorily achieved, except in an idealized non-relativistic world in which there is no creation and annihilation of particles.

The ultimate problem with this Bohmian approach is precisely the discreteness problem previously emphasized. Bohm's demonstration of the empirical equivalence of his model with the predictions of the orthodox theory depends on the idea that the experimental measuring procedure will lead to an output state that has two or more output channels that are represented by wave functions *that are non-overlapping in ordinary three-dimensional space*, and that correspond to the alternative possible observable outcomes of the measurement. But the state of a universe with no collapses at all will be one in which every physical feature of every device and every brain is completely smeared out, with no partitioning into discrete parts. The partitioning into definite distinguishable components that the founder's and von Neumann believed to require 'interventions' does not seem to emerge automatically from a universe controlled by process 2 alone. But without a partitioning into non-overlapping channels corresponding to different experiences Bohm's proof of the empirical equivalence of his approach to the orthodox theory fails.

10.3 Spontaneous-Reduction Models

One other attempted way of completing the quantum dynamics without bringing in 'the observer' is to introduce 'spontaneous reductions'. These are reductions that act according to some specified mechanical or statistical rule that does not involve consciousness, but that keeps a leash in the tendencies of the centers of large objects to become uncertain. The spontaneous reductions keep trimming back the spreading clouds so that the spread in the quantum mechanically specified locations of the (centers of the) large objects become negligible on the scale of visible objects. A model of this kind was originally proposed by Ghirardi, Rimini, and Weber, and has been pursued vigorously by Philip Pearle. The bottom line is that it has not been possible to construct a model of this sort that accommodates particle creation and annihilation and that is relativistically invariant in the same satisfactory sense that the orthodox von Neumann–Tomonaga–Schwinger theory is relativistically invariant. A quasi-relativistic theory of this kind has recently been proposed by Pearle (2005), who expounds also on the inability of these spontaneous-collapse models to do better. (See Stapp 2006b & 2007a for more details about these alternative quantum approaches to consciousness.)

11 The Basis Problem in Many-Worlds Theories

To fully appreciate the significance of the basis problem mentioned by Zurek, and of the impact of quantum decoherence on fundamental issues, one needs to understand certain subtle aspects of the connection between classical and quantum mechanics. This chapter, which is more technical than the others, explains these aspects, and, with the aid of some pictures, their relevance to the basis and decoherence problems.

11.1 Connection Between Classical Physics and Quantum Physics

Consider first a classically conceived system consisting of one single point particle confined to a large cubical box in ordinary three-dimensional space. Suppose we divide this box into a very large number N of tiny cubic regions. Then one way to represent some information about the system at some particular instant of time is to assign to each tiny cube a number 'one' or 'zero' according to whether the particle is in, or is not in, that tiny cube at that instant. Thus, at each instant, all N boxes will be assigned a 'zero' except for one box, which will be assigned a 'one'. (A special rule can be introduced to cover the case where the particle lies exactly on a boundary.) Over the course of time this 'one' will, due to the motion of the particle, occasionally jump from one tiny cubical box to an adjacent one

Information about the velocity of the particle can be added by introducing, *for each of the little coordinate-space boxes just mentioned*, a collection of M little boxes in a space that represents the velocity of the particle, or better, its *momentum*, which is the product of its velocity times its mass.

Quantum mechanics is somewhat analogous to classical *statistical* mechanics. That theory covers situations where one wishes to make statistical predictions about future observations on the basis of the known equations of motion, when one has only statistical information about the initial conditions. In this case each little box represents a tiny region in the combined coordinate–momentum space, which is called *phase space*, and the initial number assigned to this box will generally be not 'zero' or 'one', but some number in between. This number represents the initial probability that the combination of the location and the momentum of the particle lies in that tiny region. These numbers will sum to unity (one). One can let the sizes of these little boxes become increasingly small, and finally go over to a continuous 'probability density'. Then the classical equations of motion can be used to determine how this probability density changes over the course of time.

A typical 'measurement' from the classical physical-description-based point of view, is an action that answers the question: Do the position and momentum of the system at a time t lie in some specified region R in phase space? Given the initial probability conditions, the probability that the answer is 'Yes', at the time t is obtained by summing up all of the contributions to the evolved probability distribution that lie within the specified region R at the time t of the observation.

The case just described is a very simple case in which the physical system being observed is just one single point particle. But the same discussion applies essentially unchanged to any physical system, including, in particular, the brain of a conscious human being. In that case, the space in which the little boxes lie is a space each point of which represents a complete classically conceived brain, and each little box represents a tiny range of values in this space: each little box represents a tiny region in which both the location and the momentum of every particle in the brain are very close to the values specified by a classically conceived and described possible state of the brain. According to the classical conception of nature, the actual state of the person's brain at any particular instant lies in exactly one of these little boxes, and all but one box is assigned a zero. In a classically conceived *statistical* context a set of probability contributions that sums to unity can be distributed in any chosen way among these small boxes, each of which can in principle be shrunk to an arbitrarily small size.

In the quantum generalization of classical statistical mechanics the region R associated with an actual (conscious) observation cannot be represented by an arbitrarily small (or even sharply defined) region of the classically conceived phase space. The size of the – fuzzy-in-principle – region in phase space, defined in a suitable way, is a multiple of Planck's quantum of action. The intrinsic wholeness of each conscious thought renders the phase space of classical physics an inappropriate basis. The physical state of the brain is represented, rather, as a *vector* in an appropriate *vector space*, and each permissible conscious observation associated with that brain is associated with some set of mutually orthogonal (perpendicular) *basis vectors*. Thus the basic mathematical structure needed for the conscious-observation-based quantum theory of phenomena is fundamentally incompatible with the mathematical structure used in the physical-measurement-based classical theory of phenomena. An irreducible element of wholeness is present in the former but absent from the latter.

The neural correlates of our conscious thoughts are, according quantum mechanics, represented in a vector space of a very large number of dimensions. But the basic idea of a *vector in a vector space* can be illustrated by the simple example in which that space has just two dimensions.

Take a flat sheet of paper and put a point on it. (Imagine that your pencil is infinitely sharp, and can draw a true point, and perfectly straight lines of zero width.) Draw a straight line that starts at this point, called the origin, and that extends out by a certain amount in a certain direction. That directed line segment, or the displacement from the origin that it defines, is a *vector* in a two-dimensional space.

Any pair of unit-length vectors in this space that are perpendicular to each other constitute a basis in this two-dimensional space. (They are in fact an orthonormal basis, but that is the only kind of basis that will be considered here.) Because any pair of perpendicular unit-length vectors rigidly rotated by any angle between 0 and 360 degrees gives another perpendicular pair, there is an infinite number of ways to choose a basis in a two-dimensional space.

Given a basis, there is a unique way of decomposing any vector in the space into a sum of displacements, one along each of the two perpendicular basis vectors. The two individual terms in this sum are a pair of perpendicular vectors called the *components* of the vector in this basis. One such decomposition is indicated in Fig. 11.1.

If $\boldsymbol{V}$ has unit length and A and B are the lengths of the components of $\boldsymbol{V}$ that are directed along these two basis vectors, then, by virtue of the theorem of Pythagoras, $A^2 + B^2 = 1$, i.e., the sum of the two squares is unity. This is what a sum of *probabilities* should be. Consequently, the concept of probability can be naturally linked to the concept of vectors in a space of vectors. The angle Θ specifies the different observational processes that are possible in principle for vectors in this space, *and the two corresponding basis vectors correspond to*

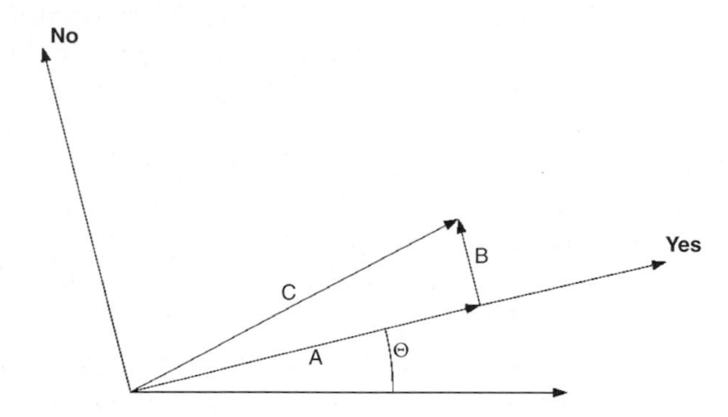

Fig. 11.1. Decomposition of vector $\boldsymbol{V}$ of length C, in a two-dimensional space, into components of lengths A and B directed along a pair of basis vectors that correspond, respectively, to the 'Yes' and 'No' answers to a possible process 1 question labeled by Θ

the two possible distinct outcomes, 'Yes' or 'No', of the observational process specified by the angle Θ.

An N-dimensional (vector) space is similar, but has N dimensions instead of just two. This means that it allows not just two mutually perpendicular basis vectors, but N of them. As a mathematical idea this is well defined. There are clearly an infinite number of ways to choose a basis – a set of mutually perpendicular unit-lenth vectors – in any space of two or more dimensions, hence an infinite set of elementary observational processes that are possible in principle. For any N, and for any basis in the N-dimensional space, there is a unique way of decomposing any vector in that space into a sum of displacements each lying along one of the mutually perpendicular basis vectors.

Each possible observational process is, according to the basic principles of quantum theory, associated with such a choice of basis vectors. The N-dimensional generalization of the theorem of Pythagoras says that the sum of the squares of lengths of the mutually perpendicular components of the unit length vector $\boldsymbol{V}$ that represents the quantum state of the physical system is unity. Consequently, the probability interpretation of the lengths of the components of the vector $\boldsymbol{V}$ carries neatly over to the N-dimensional case. Vectors in a vector space provide, therefore, a way to represent in an abstract mathematical space the probabilities associated with the perceptual realities that form the empirical basis of science.

According to quantum theory, the alternative possible phenomenal outcomes of any process of observation are associated with a set of corresponding basis vectors. Each such basis vector is associated with an – in principle fuzzy – region in the phase space of the system that is being probed, hence acted upon. This region has a prescribed size, specified by Planck's quantum of action, and only certain kinds of shapes are allowed. Thus the mathematical entities correspond possible *perceptions* in quantum theory are very restrictive as compared to the completely general sizes and shapes of the phase-space regions that are allowed to represent measurable properties of physical systems in classical physics. The transition to quantum theory imposes a severe restriction on observational realities, in comparison to the micro-structure that is deemed measurable in classical mechanics.

A quantum state of a system can be represented by a vector in a space an infinite number of dimensions. Much of von Neumann's book was devoted to the fine points of how this could be done in a mathematically well defined way. Although the number of basis vectors is infinite, it is countably infinite: the basis vectors can be placed in one-to-one correspondence to the numbers $1, 2, 3, \ldots$. That means that, *given a basis*, there is a unique decomposition of the state of the system into a *countable* set of elementary components.

The countability of the set of distinct or discrete possibilities is important. If you have a countable set of states then you could, for example, assign probability 1/2 to the first state, probability 1/4 to the second state, probability 1/8 to the third, and so on, and the total probability will add to one (unity), as a sum of probabilities should. This kind of separation into a countable set of discrete elements, each finite, is not equivalent to the separation of a continuous line into infinitesimal points: there is an element of discreteness involved with observation in quantum theory that is essentially different from what occurs in classical physics, and from what can naturally be generated from the genuinely continuous process 2 alone. The decomposition into discrete holistic components associated with a set of mutually perpendicular basis vectors in a vector space is the foundation of the relationship of the quantum mathematics to empirical phenomena. This feature blocks the association of arbitrarily tiny regions R in phase space with observation.

This discreteness aspect poses a nontrivial, and I believe fatal, difficulty for many-world theories. Scientific empirical data lies in the final analysis in our observations. But then what fixes the set of basis vectors that corresponds to some individual person's observations? Can this correspondence, which involves discreteness and wholeness and perception, be specified by the continuous micro-causal physically described process 2 alone, without any causal input from the experiential aspects of past phenomena? This is the question that lurks behind the Zurek's very true words that "Much remains to be done".

11.2 Decoherence and Discreteness in Many-Minds/Worlds Theories

An oft-repeated claim is that decoherence solves the measurement problem.

To discuss decoherence adequately it is useful to employ the density matrix formulation of quantum mechanics described by von Neumann. If the quantum mechanical representation of a system of interest, say a human brain, is represented by using a decomposition into a set of N basis states, each corresponding in principle to a possible perception, then the appropriate representation of that state considered as part of a larger universe in which it is imbedded is an $N \times N$ matrix, called the density matrix.

An $N \times N$ matrix is an array of boxes, arranged like the boxes of a crossword puzzle with N horizontal rows and N vertical columns, but with each box containing a number, instead of a letter. This matrix representation is useful when the system of interest, say a human brain, is interacting with an environment upon which no actual measurements will ever be made. In this case the observable effects of the interaction of the brain upon this empirically inaccessible environment is neatly expressed in the density-matrix representation of the brain.

The brain of an observer can be represented, then, by a matrix with a very large number N of rows and columns. The *diagonal elements* of the matrix are the elements that lie in a row and in a column that both correspond to the same basis vector. Each diagonal element can be made to correspond roughly to a smeared out classically described possible state of the entire brain (or of a macroscopic part of the brain, e.g., the frontal cortex) with the number in that box a probability associated with that possible perceptually pertinent state of the brain. The quantum generalization of classical statistical mechanics corresponds to the expansion of the diagonal line of the matrix out to the full square 'density' matrix. For any given state of the universe, the density matrix associated with the brain of interest will be filled with numbers in some particular way.

Quantum theory allows statistical predictions to be made, on the basis of the numbers in this density matrix. The off-diagonal elements control 'interference effects' between the states of the brain associated with differing row and column.

To put the idea of 'different possible actions' in a definite context, suppose you are walking in a jungle at night and a shadowy form jumps out of the darkness. The job of your brain is to evaluate your situation and construct a coordinated plan of action, perhaps to fight, or perhaps to flee. According to a classical model, your brain will, if well conditioned, decide on one plan or another, not produce both plans with no decision between them. However, in the case of a 'close call' the actual decision may depend on the particular state of the background noise associated with all of the random spikings of all the neurons in your brain.

In the quantum description there is, at the micro-level of the calcium ions entering nerve terminals a significant and unavoidable indeterminateness introduced by the narrowness of the ion channels through which the ions enter the nerve terminals. (Schwartz, 2005) Although perhaps damped out by massive parallel processing in those special cases where one particular response is overwhelmingly favored, the alternative mutually incompatible classically described possibilities of 'fight' and 'flight' could, in 'close call' cases, *both* be created and sustained by the purely physically described processing: in view of the underlying basic indeterminacy at the micro-level the Schroedinger equation could produce at the macro-level an analog in your brain of Schroedinger's cat.

Figure 11.2 shows the density matrix representation of a brain with two sets of rows singled out. The first singled-out set corresponds to brain states in which the template for action corresponding to 'fight' is active, and the other singled-out set of rows corresponds to states in which the template for action corresponding to 'flight' is active. The two corresponding sets of columns are also indicated. It is assumed that the available energy and organizational structure will go to one template or the other, so that *at the classical level of description* the two templates will not be simultaneously activated. Correspondingly, the two intervals along the diagonal corresponding 'fight' and 'flight' are well separated in Fig. 11.2. Nonzero numbers in the boxes corresponding to 'fight' rows but 'flight' columns – or vice versa – correspond to the possibility of observing interference effects between the 'fight' and 'flight' parts of the state of the brain represented by this density matrix. The diagonal elements correspond most nearly to the phase space of classical physics. However, the phase space of classical physics is not partitioned by a process – related to Planck's quantum

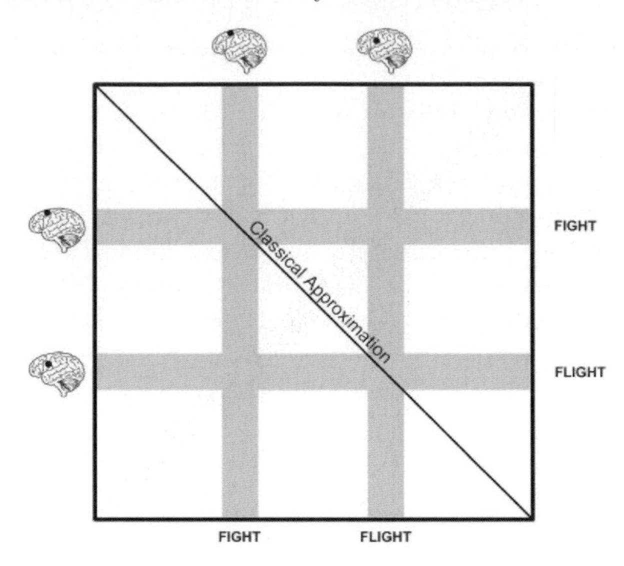

Fig. 11.2. The density-matrix representation of the brain with the sets of rows and columns corresponding to the activation of a template for a 'fight' action or for a 'flight' action both *shaded*

of action – into discrete regions of finite size and special shapes that are associated, by virtue of the workings of this process 1, with discrete, whole, alternative possible experiences.

The much-discussed decoherence effect arising from interactions with the environment is shown schematically in Fig. 11.3: the elements not lying in the shaded region are damped essentially to zero. The diagram illustrates the two main points:

1. The decoherence effect does not single out any one particular nearly classical state: it merely damps effectively to zero all significantly-non-classical possibilities, leaving the entire range of essentially classical possibilities intact and untouched, *including both the 'fight' and 'flight' portions of the (nearly classically interpretable) diagonal*. Thus environmental decoherence *produces no choice* between the (nearly) classically interpretable – but very different – states corresponding, respectively, to 'fight' or 'flight'. The entire portion of the matrix that corresponds to classically describable possibilities is retained essentially untouched.
2. The off-diagonal parts of the density matrix that can lead to interference effects between the 'fight' and 'flight' potentialities has been effectively damped out by the interaction with the environment.

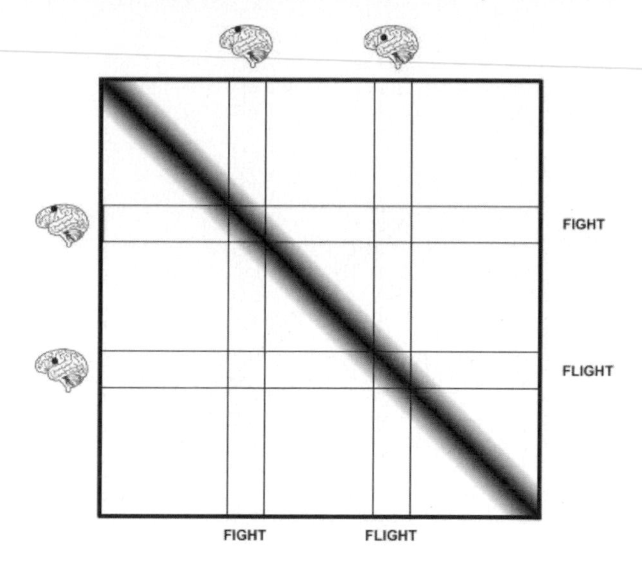

Fig. 11.3. Environmental decoherence effectively damps out the matrix elements not on the diagonal or close to them, but does not choose between alternative essentially classical possibilities

The environmental decoherence effect, being a consequence of the physically describable Schroedinger equation alone, does not come to grips with the discreteness issues pertaining to the connection of the quantum state to probabilities associated with observations. That connection involves, critically, von Neumann's process 1 intervention.

Process 1 acts, in general, upon the density matrix that specifies the state of some system that is being observed. It sets certain of the elements of that matrix to zero and leaves the rest unchanged. Figure 11.4 shows the effect on the density matrix of the particular process 1 action associated with 'fight'.

Note that process 1 is a decoherence effect, in the sense that it sets to zero certain off-diagonal elements, but leaves all diagonal elements unchanged. It is more incisive, in a certain sense, than the environmental decoherence effect in that it sets strictly to zero the elements in a region that extends right down to the (classically interpretable) diagonal. Consequently, the process 1 action carves out a set of 'Yes' diagonal elements, and, by exclusion, a complementary set of 'No' diagonal elements. The latter set consists of all the diagonal elements that are not 'Yes' elements.

It is important that the quantum decomposition into separate boxes is in terms of elements corresponding to *basis vectors associated with*

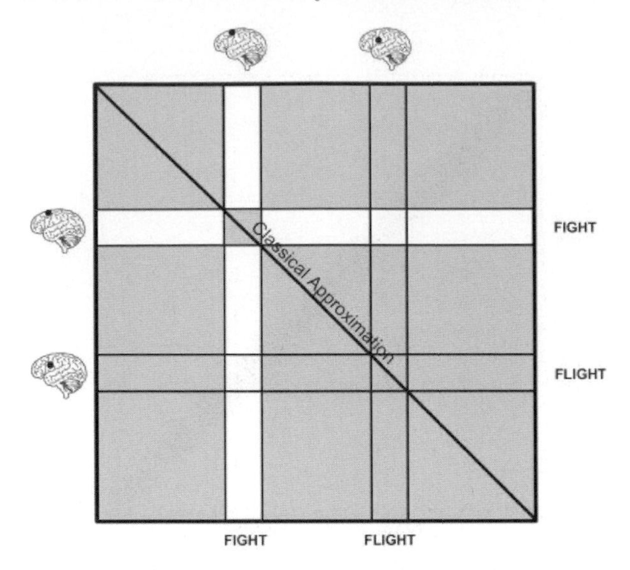

Fig. 11.4. This figure shows the effect on the density matrix representing the state of some person's brain of the process 1 action whose 'Yes' component singles out the 'fight' possibility. The process 1 action sets strictly to zero all (*whitened*) elements lying in a 'Yes' column and a 'No' row or in a 'Yes' row and a 'No' column, but leaves unchanged all other (*shaded*) elements

possible observable outcomes. It is this essential feature that establishes the connection of the quantum mathematics to empirical/phenomenal data.

Figure 11.5 shows the effect of the process 1 action shown in Fig. 11.4 upon the state of the environmentally reduced brain shown in Fig. 11.3.

Figure 11.6 shows the effect of nature's choosing the 'Yes' outcome. The surviving states of the brain are those in which the template for 'fight' action is active.

According to the precepts of quantum theory the reduction event leading to the 'Yes' state shown in Fig. 11.6 is the physically described aspect of a *psychophysical event* whose psychologically described aspect is the experiencing of the intention to perform this 'fight' action. In general, the basic realities in quantum theory are psychophysical events, and for each such event its physically described aspect is the reduction of the quantum state of an observed system to the part of that state that is compatible with the psychologically described aspect, which is an increment in knowledge entering a stream of consciousness. The evolving physical state is thereby kept in accord with our evolving

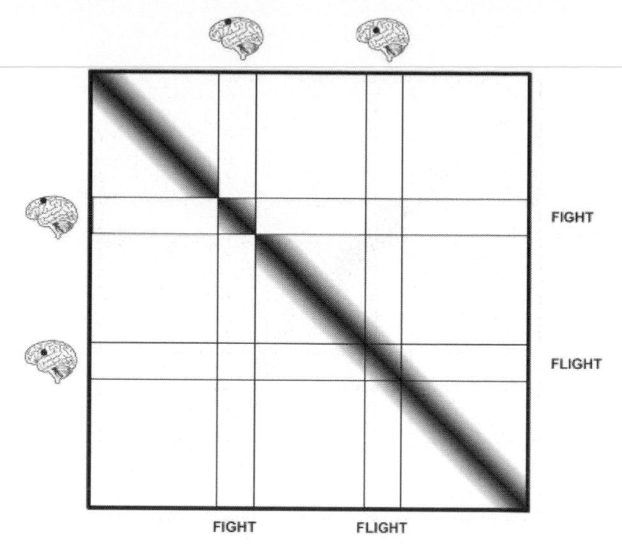

Fig. 11.5. The effect upon the environmentally reduced state of the brain produced by the process 1 action that is such that its 'Yes' outcome preserves only those states of the brain in which the 'fight' template for action is active

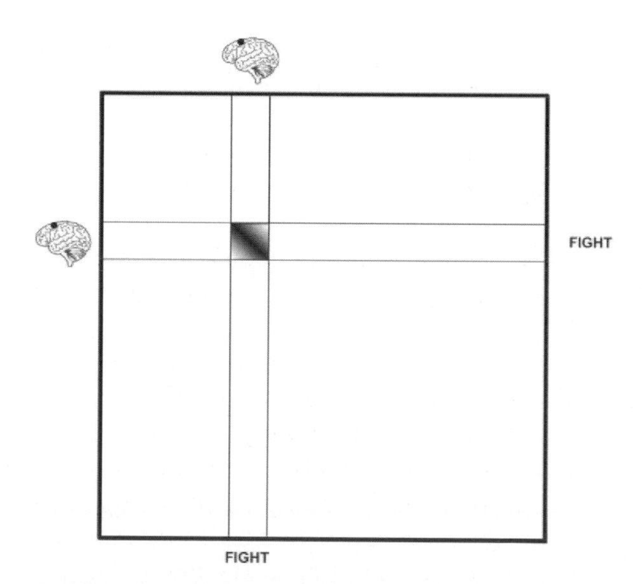

Fig. 11.6. The effect of nature's answering 'Yes' to the question: Will the template for 'fight' be active? The effect is to set to zero of all elements of the density matrix of the brain except those in the *shaded area*

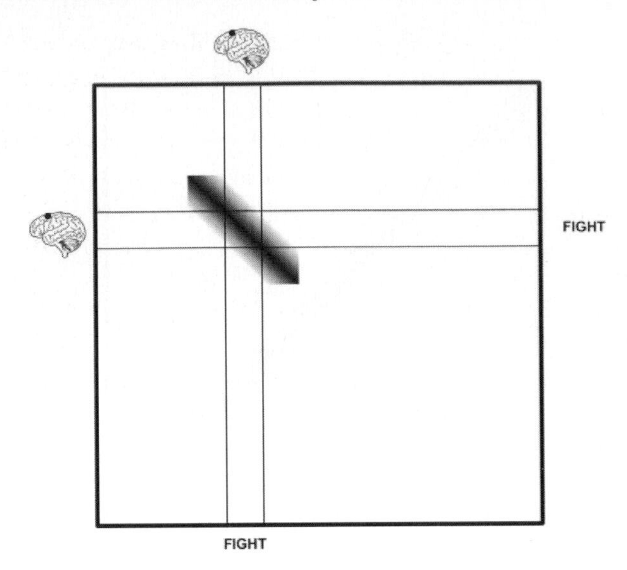

Fig. 11.7. The 'diffusion' effect generated by the normal Schroedinger equation evolution of the 'Yes' state of the brain shown in Fig. 11.6

state of knowledge, in accordance with Bohr's words cited earlier, and Heisenberg's famous statement (Heisenberg 1958, p. 100):

> The conception of objective reality of the elementary particles has thus evaporated not into the cloud of some obscure new reality concept but into the transparent clarity of a mathematics that represents no longer the behavior of particles but rather our knowledge of this behavior.

Figure 11.7 shows the effect on the 'Yes' state shown in Figure 6 that would be generated by the normal evolution in time specified by the Schroedinger equation.

The quantum zeno effect entails that if the process 1 action indicated by Fig. 11.5 is repeated sufficiently rapidly then the diffusion action indicated in Fig. 11.7 will be blocked, and the state of the brain will be restricted essentially to the 'Yes' condition, indicated in Fig. 11.6, namely to the space (set) of states such that the neurological activity identified as the template for a 'fight' action will remain activated – for longer than the classical precepts would allow. The effect of holding this template for action in place for an extended period should be to cause the intended 'fight' response to occur. In this way, anything that influences the process 1 choice of basis, and the choice of the rapidity with which the process 1 action occurs, also influences, *via the quantum laws that govern the causal connections between observation and brain activity*, the person's physical actions.

These choices are not determined by the currently known laws of physics, and they link the quantum dynamics to observation. It is therefore not unnatural that these choices should be causally affected by the phenomenal realities of the observer's stream of conscious experiences. Such a connection would speak directly to the point raised by William James, who said: "The conclusion that it [consciousness] is useful is, after all this, quite justifiable. But if it is useful it must be so through its causal efficaciousness, and the automaton-theory must succumb to common-sense" (James 1890, p. 144).

12 Despised Dualism

Scientists in different fields are free, to some extent, to use concepts that appear to work for them, without regard to other scientific disciplines. However, many of the greatest advances in science have come from unifying the treatments of neighboring realms of phenomena. We are now engaged a great scientific endeavor to rationally connect the neurophysiological and psychological aspects of the conscious brain. The problem is to understand, explain, or describe the connections between two realms that are conceived of – and are described in – two very different ways. What seems pertinent is that basic physics *was forced by the character of empirical phenomena* to an incredibly successful way to link these *same* two realms. It seems reasonable to *at least try* to apply the solution discovered by physicists to the parallel problem in neuropsychology. Why should there be such scorn in brain science for this natural and reasonable idea of bringing mind into neuropsychology in the same way that it was brought into physics in connection with the relationship between the empirically described and physically described aspects of scientific practice?

Contemporary physics is essentially psychophysical, hence dualistic. Dualism is seen as a bête noire by many philosophers. Hence the quantum approach tends to be peremptorily rejected because it belongs to this despised category. But why are dualistic theories held in such contempt? There is an historical reason.

12.1 Historical Background

I shall begin with a brief summary, abstracted from Nahmias (2002), of the principal developments in psychology during the twentieth century.

In 1898 the introspectionist E.B. Titchener delineated the proper study of psychology as the conscious mind, defined as "nothing more than the whole sum of mental processes experienced in a single lifetime". And: "We must always remember that, within the sphere of psychology, introspection is the final and only court of appeal, that psychological evidence cannot be other than introspective evidence."

However, the psychologist William James (1892), who used introspection extensively, but recognized a causal link of consciousness to brain process, lamented that psychology had not developed any laws: "We do not even know the terms between which the elementary laws would obtain if we had them."

J.B. Watson, emphasizing the failures of introspection to achieve reliable results, went to the opposite extreme. He began his 1913 behaviorist manifesto with the words: "Psychology as the behaviorist views it is a purely objective experimental branch of natural science. Its theoretical goal is the prediction and control of behavior. Introspection forms no essential part of its methods, nor is the scientific value of its data dependent upon the readiness with which they lend themselves to interpretation in terms of consciousness."

The behaviorist movement made rapid gains and in 1917 H.W. Chase wrote a summary of the year's work on *Consciousness and the Unconscious* in which he reports:

> There can be no question that consciousness is rapidly losing its standing as a respectable member of the psychologist's vocabulary. Titchener in the preface of his new book says: I have avoided the use of the word 'consciousness'. Experimental psychology has made a serious effort to give it scientific meaning, but the attempt has failed, the word is too slippery, and so is better discarded.

Technical difficulties with behaviorism began to emerge and continued to mount, but, in Nahmias's words: "It was not until Chomsky's 1959 famous review of B.F. Skinner's *Verbal Behavior* that the tide fully turned against trying to treat language, including reports about human experience, just like any other behavior." This turning of the tide meant that behaviorism failed precisely for the point at issue: the connection of physical process to conscious process. Yet the pariah status assigned to dualism by behaviorists lingered on after the fall of behaviorism, and it still persists today. But why should this bias continue after the demise of the discredited philosophy that spawned it?

12.2 A Flawed Argument

Daniel Dennett (1991) gives a reason. His book *Consciousness Explained* has a chapter entitled *Why Dualism Is Forlorn*, which begins with the words:

> The idea of mind as distinct [...] from the brain, composed not of ordinary matter but of some other special kind of stuff is dualism, and it is deservedly in disrepute today. [...] The prevailing wisdom, variously expressed and argued for is materialism: there is one sort of stuff, namely matter – the physical stuff of physics, chemistry, and physiology – and the mind is somehow nothing but a physical phenomenon. In short, the mind is the brain.

Dennett then asks: "What, then, is so wrong with dualism? Why is it in such disfavor?" He answers:

> A fundamental principle of physics is that any change in the trajectory of a particle is an acceleration requiring the expenditure of energy [...] this principle of conservation of energy [...] is apparently violated by dualism. This confrontation between standard physics and dualism has been endlessly discussed since Descartes' own day, and is widely regarded as the inescapable flaw in dualism.

This argument depends on identifying 'standard physics' with classical physics. The argument collapses when one goes over to contemporary physics, in which, due to the Heisenberg uncertainty principle, trajectories of particles are replaced by cloud-like structures, and in which conscious choices can influence physically described activity without violating the conservation laws or any other laws of quantum physics. *Contemporary physical theory allows, and its orthodox von Neumann form entails, an interactive dualism that is fully in accord with all the laws of physics*. Any perception merely reduces the possibilities.

12.3 Squaring with Contemporary Neuroscience

How does the quantum conception of mind–brain dynamics square with contemporary neuroscience?

Steven Pinker is an able reporter on contemporary neuroscience. In the lead article *The Mystery of Consciousness* in the January 29, 2007

Mind & Body Special Issue of Time Magazine he notes that while certain mysteries remain, neuroscientist agree on one thing: "Francis Crick called it 'the astonishing hypothesis' – the idea that our thoughts, sensations, joys and aches consist entirely of physiological activity in the tissues of the brain."

Of course, the phrase 'physiological activity' needs to be replaced by 'psychophysiological activity' since this activity is being explicitly asserted to have psychological or experiential content. Later Pinker says that: "Consciousness turns out to consist of a maelstrom of events distributed across the brain." These events should evidently be labeled psychophysical events, since being located in the brain is a physical attribute, while being the components of consciousness entails that these events have psychological aspects.

These psychophysiological or psychophysical characterizations fit quantum theory perfectly. According to von Neumann's formulation each of the quantum events in the brain has both a psychological aspect and a physical aspect. The physical aspect is the jump of the quantum state of the brain to that part of itself that is compatible with the increment in knowledge specified by its psychologically described aspect. It is this tight linkage between the psychologically and physically described aspects of the events that keeps a person's brain in alignment with his or her experiences. These repeated reductions are both possible and needed because the indeterminacy present at the microscopic/ionic level, keeps generating at the macroscopic level a profusion of brain states corresponding to mutually incompatible observations. These dynamically needed interventions, whose causal origin is left unspecified by the physical theory, provide a natural vehicle for mental causation.

This all depends on accepting the utility of the quantum mechanical program of building science's conception of nature on the notion of a sequence of macroscopically localized psychophysical events, rather than on the notion of mindless matter.

Pinker refers to 'The Hard Problem'. He says:

> The Hard Problem is explaining how subjective experience arises from neural computation. The problem is hard because no one knows what a solution would look like or even is a genuine scientific problem in the first place. And not surprisingly everyone agrees that the hard problem (if it is a problem) is a mystery.

Of course this 'hard' problem is – and will remain – a mystery insofar as one's thinking is imprisoned within the fundamentally invalid conceptual framework postulated by classical physics, which has no rational place for consciousness. Within that framework the problem is seen to be "explaining how subjective experience arises from neural computation", since all that is given is mindless matter. But the mystery immediately dissolves when one passes over to quantum theory, which was formulated from the outset as a theory of the interplay between physical descriptions and conscious thoughts, and which comes with an elaborate and highly tested machinery for relating these two kinds of elements.

Some quantum physicists want to justify basing neuroscience on classical physics by suggesting that once the neural activity reaches a classically describable level, say at the firing of a neuron (i.e., the triggering of an action potential), one may assume that the quantum jump from 'potential' to 'actual' has occurred, and hence that one can deal simply with the actualities of neuron firings, and ignore their quantum underpinnings.

That approach would be a misuse of the quantum mechanical use of the concepts of classical mechanics. The founders of quantum mechanics were very clear about the use, in the theory of observations, of the concepts of classical mechanics. Those concepts were needed and used in order "to communicate to others what we have done and what we have learned". The use of the classical concepts is appropriate in that context because those pertinent experiences are actually describable in terms of the classical concepts, not because something was mysteriously supposed to actually happen merely when things became big enough for classical ideas to make sense. That criterion was too vague and ambiguous to be used to construct a satisfactory physical theory. The boundary between the large and the small could be shifted at will, within limits, but actuality cannot be shifted in this way.

When one is describing one's perceptions of devices lying outside one's body the experience itself is well described in terms of classical ideas about where the parts of the device are and how they are moving. But one's subjective phenomenal experience is not geometrically similar to the pattern of neural firings that constitute the neural correlate of that experience.

If one assumes that the reduction events in the subject's brains are tied fundamentally to classicality per se, rather than to increments in the subject's knowledge, then one loses the essential connection between physical description and subjective experience that quantum theory is designed to provide This quasi-classical approach of accepting quantum mechanics at the microscopic level, but tying the reduction events occurring in the subject's brain to some objective condition of classicality, rather than to the subject's experiences, has the great virtue – relative to the approach of simply accepting a fully classical conception of the brain – of not just ignoring a hundred years of development in physics. However, in the context of solving the problem of the mind–brain connection, it inherits the fatal deficiency of the classical approach: the conceptual framework does not involve mind. There is, as in the classical approach, no intrinsic conceptual place for, or dynamical need for, our conscious experiences. There is within the given structure no entailment either of any reason for conscious experiences to exist at all, or of any principle that governs how these experiences are tied to brain activity. "The Hard Problem of explaining how subjective experience arises from neural computation" remains, as Pinker said "a mystery". Moreover, the quasi-classical approach inherits also the principal difficulty of all the quantum theories that accept reductions, but reject the orthodox principle of placing the reduction events at the boundary between the physically described and psychologically defined aspects of our scientific understanding of nature. Where, within such an approach that does not involve consciousness, can one find either any reason for any reduction to occur at all, or any objective principle that specifies where, between one single atom and the more than 10^{24} atoms in the brain, do the collapses occur. Orthodox quantum theory ties these two problems of 'consciousness' and 'collapse' together in a practically useful way, and provides, simultaneously, a way for the universe to acquire meaning.

13 Whiteheadian Quantum Ontology

Upon completing my article entitled *The Copenhagen Interpretation* (Stapp 1972), I sent the manuscript to Heisenberg for his approval or reaction. He expressed general approval, but raised one point:

> There is one problem I would like to mention, not in order to criticize the wording of your paper, but for inducing you to more investigation of this special point, which however is a very deep and old philosophical problem. When you speak about the ideas (especially in [section 3.4]) you always speak of human ideas, and the question arises, do these ideas 'exist' outside of the human mind or only in the human mind? In other words: have these ideas existed at the time when no human mind existed in the world?

He continued:

> I am enclosing the English translation of a passage in one of my lectures in which I have tried to describe the philosophy of Plato with regard to this point. The English translation was done by an American philosopher who, as I think, uses the philosophical nomenclature correctly. Perhaps we could connect this Platonic idea with pragmatism by saying: It is 'convenient' to consider the ideas as existing outside the human mind because otherwise it would be difficult to speak about the world before human minds have existed.

These remarks highlight the fact that in the foregoing chapters I have adhered to the Copenhagen pragmatic stance of erecting science upon human knowledge. Yet science encompasses cosmology, and also our attempts to understand the evolutionary process that created our species. If we want to address the basic question of the nature of human beings then we need more than merely a framework of practical rules that work for us. We need to be able to see this pragmatic anthropocentric theory as a useful distillation from an underlying non-anthropocentric ontological structure that places the evolution of our conscious species within the broader context of the structure of nature herself. We need a fundamentally non-anthropocentric ontology within which the anthropocentric pragmatic theory is naturally imbedded.

That is a big order! Fortunately, however, there already exists such an ontology. It is the ontology proposed by Alfred North Whitehead, which beautifully accommodates orthodox pragmatic quantum theory.

I recently gave some lectures and published a paper on this topic, and will, for the rest of this chapter, follow closely that text (Stapp 2007b). This brings in certain overlaps with things that have already been said here, but that warrant re-saying in this broader context.

> Nature and Nature's Laws lay hid at night
> God said 'Let Newton be!' and all was light. (Alexander Pope)

> In our description of nature the purpose is not to disclose the real essence of phenomena, but only to track down as far as possible relations between the multifold aspects of our experience. (Niels Bohr)

These two quotations highlight the question: What is the proper task of science? Is it to illuminate the nature of reality itself, as Alexander Pope proclaimed was already achieved by Isaac Newton? Or should the goal of science be curtailed in the way recommended by Niels Bohr?

Bohr (1958, p. 71) asserted that:

> [...] the formalism does not allow pictorial representation along accustomed lines, but aims directly at establishing relations between observations obtained under well-defined conditions.

However, the impossibility of representing reality along accustomed lines does not automatically preclude every kind of conceptualization. Perhaps an uncustomary idea will work. Even Newton's mechanical conception was not customary when he proposed it. Hence if advances in science reveal the incompatibility of the empirical evidence with customary pictorial representations then perhaps the construction of a new vision of reality is needed, instead of meek resignation to the construction of practically useful rules.

To operate most effectively in the physical world one needs an adequate conception of oneself operating within that world and upon it. Optimal functioning is impaired if you are armed only with blind computational rules, severed from a rationally coherent conception of yourself applying those rules.

There is, of course, no guarantee that our species can come up with an adequate conceptualization of our mindful selves acting in and upon the world. And even if such a conceptualization were uncovered, there is no assurance that it is unique. However, neither the fear of failure nor the specter of non-uniqueness constitutes a sufficient reason to refrain from at least trying to find some satisfactory understanding of our conscious selves imbedded in the reality that surrounds and sustains us.

Due undoubtedly, at least in part, to the impact of Bohr's advice, most quantum physicists have been reluctant even to try to construct an ontology compatible with the validity of the massively validated pragmatic quantum rules involving our causally efficacious conscious thoughts. However, due to this reticence on the part of quantum physicists we are faced today with the spectacle of our society being built increasingly upon a conception of reality erected upon a mechanistic conception of nature now known to be fundamentally false. Specifically, the quintessential role of our conscious choices in contemporary physical theory and practice is being systematically ignored and even denied. Influential philosophers, pretending to speak for science, claim, on the basis of a grotesquely inadequate old scientific theory, that the (empirically manifest) influence of our conscious efforts upon our bodily actions, which constitutes both the rational and the intuitive basis of our functioning in this world, is an illusion. As a consequence of this widely disseminated misinformation the 'well informed' officials, administrators, legislators, judges, and educators who actually guide the development of our society tend to direct the structure of our lives in ways predicated on false premises about 'nature and nature's laws'.

Bohr's pragmatic quantum philosophy emphasizes the active role that we human beings play in the development of our scientific knowledge. But this approach can easily lead to an anthropocentric conception of reality.

A rational escape from this parochialism is provided by work of the eminent philosopher, physicist, and logician Alfred North Whitehead. He created a conception of natural process that captures the essential innovations wrought by quantum theory in a way that allows the human involvement specified by quantum theory to be understood within a fundamentally non-anthropocentric conception of nature as a whole.

Whitehead struggled to reconcile the findings of early twentieth century physics with the insights and arguments of the giants of Western philosophy, including, most prominently, Plato, Aristotle, Descartes, Leibniz, Locke, Hume, Kant, and William James. But although Whitehead had the hints about "abrupt quantum jumps" and "objective tendencies for these jumps to occur" that came from early quantum theory, and was familiar with Einstein's special and general theories of relativity, he was not acquainted with the important and sophisticated developments in relativistic quantum field theory represented by the mid-century works of Tomonaga and Schwinger.

I shall describe here a conception of reality that stems primarily from the ontological ideas of Werner Heisenberg, the principal founder of quantum theory, expressed within an ontological construal of von Neumann's formulation, as revised by Tomonaga and Schwinger to form the foundation of relativistic quantum theory. This relativistic quantum ontology is in close accord with many key ideas used by Whitehead.

It will both clarify this quantum ontology and bring it into a certain correspondence with the Whiteheadian framework to begin by quoting Whitehead's clear enunciations of those key ideas. On the other hand, I make no claim to encompass every pronouncement of Whitehead, who wrote long before the work of Tomonaga and Schwinger. Indeed, I shall always take the quantum theoretical findings as preeminent, and use only those assertions of Whitehead that mesh nicely with, and flesh out, the ontological construal of the quantum formalism that springs naturally from the formulation of John von Neumann, as brought into accord with the precepts of the special theory of relativity by the works of Tomonaga and of Schwinger.

The core issue for both Whiteheadian process and quantum process is the emergence of the discrete from the continuous. This problem is illustrated by the decay of a radioactive isotope located at the center of a spherical array of a finite set of detectors, arranged so that they cover the entire spherical surface. The quantum state of the positron emitted from the radioactive decay will be a continuous spherical wave, which will spread out continuously from the center and eventually reach the spherical array of detectors. But only one of these detectors will fire. The total space of possibilities has been partitioned into a discrete set of subsets, and the prior continuum is suddenly reduced to some particular one of the elements of the selected partition.

But what fixes, or determines, this particular partitioning of the continuous whole into the discrete set of subsets?

The orthodox answer is that the experimenter decides.

Yet if the experimenter himself is made wholly out of physical particles and fields then his quantum representation by a wave function must also be a continuous function. But how can a smeared out continuum of classically conceivable possibilities be partitioned into a set of discrete components by an agent who is himself a continuous smear of possibilities. How can the definite fixed boundaries between the discrete elements of the partition emerge from a continuous quantum smear?

None of the founders of quantum theory could figure out how this could happen – nor has anyone since. Von Neumann, in his rigorous formulation of the mathematics of quantum theory, calls this partitioning action an 'intervention': it is an intervention into the continuous deterministic Schroedinger-equation-controlled evolution of the physically described aspects of the universe.

This 'discreteness' problem is resolved in orthodox quantum theory, and in actual scientific practice, by what Heisenberg and Bohr call "a choice on the part of the experimenter". Von Neumann calls the manifestation in the physical world of this choice by the name 'process 1'. I shall call by the name 'process zero', the process that *selects* the particular partitioning specified by the physically described process 1.

What seems clear is that this partitioning cannot arise from the physically described aspects of the world acting alone: continuous smears acting in accord with the smoothing Schroedinger equation cannot create a discrete partitioning in a finite time. However, the experimenter feels that his consciousness is playing a role. Indeed, if the physically described aspects alone cannot do the job, and it feels like consciousness is helping, then why not try that idea out? Consciousness is, after all, the only other thing in our ontological arsenal.

But how, then, can we then understand, coherently and rationally, how to make that idea work?

The plan of the presentation is this:

1. Specify by using Whitehead's own words what I take to be his key ideas.
2. Put them coherently together to form the spacetime aspects of Whiteheadian process.
3. Describe the basic structure of ontologically conceived Tomonaga–Schwinger relativistic quantum field theory.
4. Put these elements coherently together to form the spacetime picture of quantum process.
5. Note the identity of these two spacetime pictures.
6. Note some further identities, and propose a unified non-anthropocentric Whiteheadian quantum ontology.

This ontology is still not completely specified. But it is far more structured than a general pan-psychism. It specifies distinctive conditions pertaining to space, time, causation, the notion of the 'now', the physically and psychologically described aspects of nature, and the nature of conscious agents. The empirically validated anthropocentric concepts of contemporary orthodox pragmatic quantum theory have thereby become imbedded in a general non-anthropocentric theory of reality.

13.1 Some Key Elements of Whitehead's Process Ontology

I shall now state what I take to be Whitehead's key principles, expressed in Whitehead's own words, taken from his book *Process and Reality*.

Whitehead's first principle is that the world is built out of actual entities/occasions:

> 'Actual entities' – also termed 'actual occasions', are the final real things of which the world is made. (PR p. 18)

> The final facts are, all alike, actual entities, and these actual entities are drops of experience, complex and interdependent. (PR p. 18)

Whitehead accepts James's claim about the droplike (atomic or indivisible) character of experience:

> Either your experience is of no content, of no change, or it is of a perceptible amount of content or change. Your acquaintance with reality grows literally by buds or drops of perception. Intellectually and on reflection you can divide them into components, but as immediately given they come totally or not at all. (W. James 1911)

Whitehead builds also upon James's claim that: "The thought is itself the thinker":

> If the passing thought be the directly verifiable existent, which no school has hitherto doubted it to be, then that thought is itself the thinker, and psychology need not look beyond. (James 1890, p. 401)

Thus the 'actual entities' are the 'drops of experience' themselves, not some soul-like entities that know them. Your awareness of your 'self' must be an aspect of your thoughts, and there is no rational need for, additionally, something besides or beyond the reality that is that awareness itself.

Whitehead draws a basic distinction between the two kinds of realities upon which his ontology is based: 'continuous potentialities' versus 'atomic actualities':

> Continuity concerns what is potential, whereas actuality is incurably discrete. (PR p. 61)

Another Whiteheadian precept is that actual entities decide things!

> Actual entities [...] make real what was antecedently merely potential. (PR p. 72)

> Every decision is referred to one or more actual entities [...]. Actuality is decision amid potentiality. (PR p. 43)

> Actual entities are the only reasons. (PR p. 24)

Another of Whitehead's key ideas is that each (temporal) actual entity is associated with a region of space:

> Every actual entity in the temporal world is to be credited with a spatial volume for its perspective standpoint [...]. (PR p. 68)

A closely associated idea is that these regions 'atomize' spacetime:

> The actual entities atomize the extensive continuum. This [spacetime] continuum is in itself merely potentiality for division. (PR p. 67)

> The contemporary world is in fact divided and atomic, being a multiplicity of definite actual entities. These contemporary actual entities are divided from each other, and are not themselves divisible into other contemporary actual entities. (PR p. 62)

The central idea in Whitehead's philosophy is his notion of process:

> The many become one, and are increased by one. (PR p. 21)

Thus in Whiteheadian process the world of fixed and settled facts grows via a sequence actual occasions. The past actualities generate potentialities for the next actual occasion, which specifies a new spacetime standpoint (region) from which the potentialities created by the

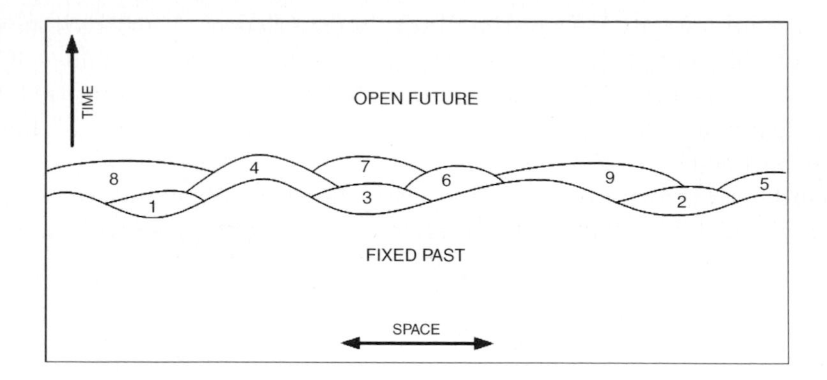

Fig. 13.1. A representation of the spacetime aspects of the Whiteheadian process of creation

past actualities will be prehended (grasped) by the current occasion. This basic autogenetic process creates the new actual entity, which, upon its creation, contributes to the potentialities for the succeeding actual occasions.

Nature's process assigns a separate spacetime region to each actual entity, and this process fills up, step-by-step, the spacetime region lying in the past of the advancing sequence of spacelike surfaces 'now', as indicated in Fig. 13.1.

The bottom curvy line represents the (spacelike) three-dimensional surface 'now' that separates – at some stage of the process of creation – the spacetime region corresponding to the fixed and settled past from the region corresponding to the potential future. Each new actual occasion has a standpoint spacetime region, which gets added to the past, thereby pushing slightly forward the boundary surface 'now'. The small regions with numbers indicate the standpoints of a succession of actual occasions each representing a step in the creative process.

This conception of a growing actual spacetime region – filled with (the standpoints of) the growing set of past actual occasions – that advances into the strictly potential open future, constitutes a certain resolution to the famous debate between Newton and Leibniz about the nature of space. Newton's conception, described in the Scholium in his main work, *Principia Mathematica*, was essentially a *receptacle* conception, in which space is an empty container into which movable physical objects can be placed.

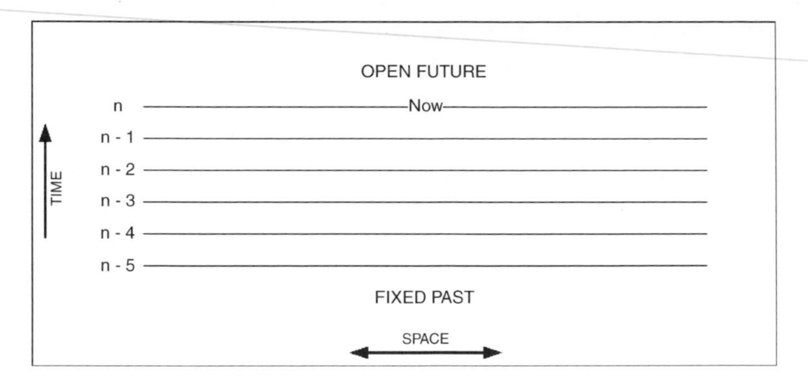

Fig. 13.2. A representation of the spacetime structure in non-relativistic quantum theory. At each one of a sequence of constant-time surfaces an 'intervention' occurs in association with an abrupt jump to a new quantum state $\Psi(t)$

Leibniz argued for the *relational* view that space is naught but relations among actually existing entities: completely empty space is a nonsensical idea.

But Whitehead's *actual* spacetime is filled by actual atomic (indivisible) entities. Thus it is not empty. On the other hand, there is also a yet-to-be-filled spacetime future, which, however, is still a mere potentiality.

This Whiteheadian idea of the growing 'past' can be contrasted with the corresponding idea in non-relativistic quantum physics. In non-relativistic quantum physics the growing 'past' lies behind an advancing (into the future) sequence of constant-time instants 'now', as illustrated in Fig. 13.2.

In non-relativistic quantum theory (NRQT) the fixed past advances into the open future in layer-cake fashion, one temporal layer at a time. Each quantum reduction event occurs at some particular time NOW, *but over all of space*. In von Neumann's nonrelativistic quantum theory this event produces the new quantum state $\Psi(t)$ of the universe at the instant labeled by the time t.

This non-relativistic spacetime structure is replaced in Tomonaga–Schwinger relativistic quantum field theory (RQFT) by a different structure.

13.2 From von Neumann NRQT to Tomonaga–Schwinger RQFT

In RQFT the NRQT state $\Psi(t)$ is replaced by $\Psi(\sigma)$. Here t specifies a continuous three-dimensional surface in the four-dimensional spacetime continuum, with all spatial points lying at the same time t. But σ specifies a continuous three-dimensional surface in the four-dimensional spacetime continuum, with every point on that surface spacelike-separated from every other point (i.e., no point on the surface can be reached from any other point by moving at the speed of light or slower).

The Whiteheadian spacetime structure represented in Fig. 13.1 represents also the spacetime structure of a sequence of discrete actualization events in the Tomonaga–Schwinger formulation of relativistic quantum field theory. In this case, the sequence of spacelike surfaces 'now' represents the relativistic generalizations of the sequence of fixed-time surfaces upon which, in the non-relativistic formulation of quantum theory, the quantum state (of the universe) is (re)defined just after a quantum jump.

In the relativistic case the bottom wavy line in Fig. 13.1 represents some initial surface σ, an initial NOW. In the dynamical evolution of the quantum state this surface pushes continuously forward first though the spacetime region labeled 1. This unitary evolution, via the relativistic generalization of the Schroedinger equation, leaves undisturbed the aspects of the state $\Psi(\sigma)$ associated with the rest of the initial surface σ.

Then a new quantum 'reduction' event occurs. It acts *directly* (via projection) only on the new part of the surface, the part represented by the top boundary of region 1. But this direct change causes *indirect* changes along the rest of the surface σ, due to quantum entanglements. These 'indirect changes' produce the 'faster-than-light' effects called by Einstein 'spooky actions at a distance' (see Appendices E–G).

The evolutionary process then advances the surface NOW next through region 2, then through region 3, etc. After each successive advance into the future, a quantum reduction event occurs. It is associated with a certain mathematical 'projection' operator that acts directly only on the new part of the current surface NOW, but indirectly (via entanglement) on the entire surface NOW, at least in principle.

13.3 Similarities Between Whitehead's Ontology and Ontologically Construed RQFT

Notice the identity, as regards the spacetime development indicated in Fig. 13.1, of the RQFT and the Whitehead ontologies.

But there are further correspondences. The first concerns the matching of the Whiteheadian connections between 'objective potentia' and 'subjective knowledge' with those of the Heisenberg's quantum ontology. According to Heisenberg (1958b, p. 53):

> The probability function combines objective and subjective elements. It contains statements about possibilities or better tendencies ('potentia' in Aristotelian philosophy), and these are completely objective, [...] and it contains statements about our knowledge of the system, which of course are subjective in so far as they may be different for different observers.

13.4 The Transition from 'Potentiality' to 'Actuality' in Quantum Mechanics

Heisenberg (1958b, p. 54):

> [...] the transition from the 'possible' to the 'actual' takes place during the act of observation.
>
> The observation itself changes the probability function discontinuously; it selects of all possible events the actual one that has taken place. Since through the observation our knowledge of the system has changed discontinuously, its mathematical representation has also undergone the discontinuous change and we may speak of a 'quantum jump'.

13.5 Compatibility with Einstein's (Special) Theory of Relativity

Within Tomonaga–Schwinger RQFT all *predictions* are independent of the sequential ordering of spacelike separated events, e.g., switching the sequential orderings of the occasions labeled 1 and 2 in Fig. 13.1 changes no prediction of the theory.

Furthermore, no 'signal' (controlled message) can be transmitted faster than the speed of light.

Quantum theory is designed to be a theory of predictions, and the predictions of RQFT conform to the demands of Einstein's (special) theory of relativity: the predictions do not depend upon which one of any two spacelike separated events occurs first in the sequential unfolding of actuality. Furthermore, by virtue of the detailed structure of the quantum rules, the indirect effect, via entanglement, of a quantum event occurring in one region upon predictions/potentialities pertaining to a faraway (spacelike separated) region cannot be used to transmit a 'signal' (a controllable message) faster than the speed of light.

13.6 The Psychophysical Building Blocks of Reality

In the Whiteheadian ontologicalization of quantum theory, each quantum reduction event is identified with a Whiteheadian actual entity/occasion.

Each Whiteheadian actual occasion/entity has a 'mental pole' and a 'physical pole'.

There are two kinds of actual occasions. Each actual occasion of the first kind is an intentional probing action that *partitions* a continuum into a collection of discrete experientially different possibilities. Each actual occasion of the second kind selects (actualizes) one of these discrete possibilities, and obliterates the rest.

According to this Whiteheadian quantum ontology, objective and absolute actuality consist of a sequence of psychophysical quantum reduction events, identified as Whiteheadian actual entities/occasions.

These happenings combine to create a growing 'past' of fixed and settled 'facts'.

Each 'fact' is specified by an actual occasion/entity that has both a physical aspect (pole) and a mental aspect (pole), and a region in spacetime from which it views reality. I take the physical pole or aspect of the actual occasion to consist of a physically/mathematically described *input* and a physically/mathematically described *output*. The physical input (output) is precisely the part of the physically described state of the universe that is localized, just before (after) the jump, on the *front* boundary of the standpoint region associated with the actual occasion.

The mental pole also consists of an input and an output. The mental inputs and outputs have the ontological character of thoughts, ideas, or feelings. The mental inputs are drawn primarily from the mental outputs of the prior occasions, and the mental output of the current occasion is the bud of experience created by/at this current event or occasion.

The process by which the mental and physical inputs are combined to produce mental and physical outputs involves, according to Whitehead, aspects identified as appetites, evaluations, and satisfactions. Thus idea-like qualities are asserted to enter into the dynamics of the basic process that creates the actual occasions, and hence reality itself.

The paradigmatic example of an actual occasion is an event whose mental output is an addition to a human stream of conscious events, and whose physical output is the actualized neural correlate of that mental output. Such events are 'high-grade' actual occasions. But the Whitehead quantum ontology postulates that simpler occasions that have lower-grade outputs also exist. Thus the Whitehead quantum ontology is essentially an ontologicalization of the structure of orthodox relativistic quantum field theory, stripped of any anthropocentric trappings, but supplied with an internal creative process that makes ideas dynamically effective. This approach takes the physically described and psychologically described aspects of contemporary orthodox relativistic quantum field theory to be exemplars of the elements of a general non-anthropocentric ontology.

This putative understanding of the way nature works is only an outline, the details of which can be filled in when new more incisive data that need to be accommodated become available. The theory is not *implied* by the currently available empirical data, but it gives a rationally coherent way to accommodate:

1. "the element of wholeness [...] completely foreign to the classical physical principles" noted by Bohr,
2. the "buds of perception" noted by James,
3. the concordance with findings of Western philosophy discussed by Whitehead.

This effort to ontologicalize the anthropocentric, pragmatic, orthodox quantum mechanics of its founders, and of von Neumann, may seem misdirected. For how does this observation-dependent theory apply to the formation of a track in a cloud chamber? The physical happenings in the chamber *seem* to have, fundamentally, very little to do with any act of observation: our human involvement seems only incidental. Some physicists want therefore to conclude that the collapse events in cloud chambers are instigated by purely physical causes, and that this conclusion probably holds also for brain events as well. But I believe the founders reflected more profoundly about these matters than many of those who followed.

Classical intuition indeed suggests that a well defined cloud chamber track comes into existence without an involvement of a mental input of any kind. Yet if it be granted that the coming into being of a particular track is a quantum event, which needs to be described not in terms of classical phase space but in terms of quantum concepts, namely in terms of vectors in a vector space, and a choice of basis vectors, then the problem of what chooses the basis must be dealt with in some way. The suggestion here is that, counter-intuitively, the entry into actuality of the element of discreteness/wholeness represented by von Neumann's process 1 action cannot be adequately represented within the conceptual framework of classical physics. The proposal is not that every quantum event need be associated with a reality of exactly the kind that populate our human streams of consciousness. It is rather that every quantum event is associated with an element that cannot be adequately conceptualized in terms of the precepts of classical physics, but that resides a realm of realities that are not describable in terms of the concepts of classical physics, but that *include* our conscious thoughts, ideas, and feelings.

Von Neumann's analysis of measurements shows that for all practical purposes one can assume that an appearing track actually came into being without any dependence upon our observations of it. But some sort of intervention is then needed, and a natural possibility is that any actual intervention is *formally* like an actual human observation.

This formal similarity to a human intervention means not only the occurrence of the needed choice of a basis in a vector space – a choice that injects an "element of wholeness completely foreign to classical physical principles" – but also the resolution of the uncertainty-principle-generated ambiguities by the imposition of a conceptual element, in accordance with Whitehead's demand for a mental pole. This conceptual component can have no more complexity than the physical structure that will, after the event, represent it in the quantum physical state. The human exemplars of an event have the complex marching-band structure described in Sect. 6.6 of Stapp (1993/2004a), which accounts for the senses of 'duration', 'knowing', and 'belonging to a stream of consciousness' characteristic of a human thought.

14 Interview

At the end of the summer of 2006 Harald Atmanspacher conducted an interview of me that appeared in the September 2006 issue of J. Consciousness Studies (Volume 13, No. 6). Professor Atmanspacher raised many pertinent questions that had not been dealt with in my prior writings, and have not been adequately covered in the foregoing parts of this book. My answers added important details to my elaboration of von Neumann's work. Atmanspacher's formulations of his questions have been widely praised, and any attempt by me to re-structure the content of the interview would be inappropriate. I shall therefore, with his permission, and that of JCS, reproduce that interview here:

HA: You have been actively interested in the relationship between mind and matter for almost half a century. Shortly after receiving your PhD at Berkeley, you went to work with Wolfgang Pauli at the ETH in Zurich, in 1958, the year Pauli died. During that period, you told me, you drafted a manuscript entitled *Mind, Matter and Quantum Mechanics*, which was never published. But its title reappeared in your book of 1993. What stimulated your interest so early on in your career, and what were your ideas at that time?

HPS: 1959 was indeed early in my career as a PhD, but more than a dozen years into my concerns with these matters. Already in high school I had become very interested in the wave–particle puzzle, and my driving motive in becoming a physicist was really to solve that mystery. Looking now at my 1959 essay I find it remarkably mature. I had a solid grasp of the technical and philosophical aspects of the situation. I find in it today nothing that I would emend or consider naive or deficient. It is a well-reasoned and sober assessment of the situation, and ends with the conclusion that quantum theory "primarily is a synthesis of the idealistic and materialistic world views. To some extent it also reconciles the monistic and pluralistic attitudes, provides a natural understanding of creation, and permits a reconciliation of the deterministic aspects of nature with the action of free will." I now say much more about these matters, but nothing contrary to what I said then.

HA: Since a bit more than a decade, the problem of how to relate consciousness to brain activity has been put back onto the agenda, first in the philosophy of mind, notably due to the courageous efforts of David Chalmers and others. This has led to an increased attention in other fields as well, including cognitive neuroscience, complex systems research, evolutionary biology, and others. However, I think it is fair to say that the mainstream position in the sciences is still that mental activity can be reduced to brain activity in the sense that the mind will be completely understood once the brain is completely understood. Yet there are counterarguments against this position, for instance the famous qualia arguments. How do you think about them, and which of these counterarguments appear to be most striking to you?

HPS: I believe that the arguments advanced in favor of the idea that 'understanding the brain' entails 'understanding the mind' are malformed and irrational. What does 'understanding the brain' mean? What does the word 'brain' mean as opposed to 'mind'? The aimed-at, and completely reasonable, meaning in this context of the phrase 'understanding the *brain*' is that this understanding should be basically in terms of the laws and concepts of *physics*. If 'understanding the brain' is not basically tied into understanding the brain in terms of the laws and concepts of physics then the notions 'mind' and 'brain' are nebulous and ill-defined, and no sharp conclusions can be reached. But if the phrase means understanding the brain in terms of the laws and concepts of physics then the first question is: *which physics, classical or quantum?*

The answer is clear! The classical laws are fundamentally incorrect at the ionic level at which the basic dynamics occurs, hence one must in principle use the quantum laws and concepts. There is no rational controversy about whether or not quantum effects occur in the brain – of course they do! The crucial question is the extent to which the quantum, as opposed to classical, precepts are essential for the dynamics of the brain; and to what extent a classical approximation is valid in a warm, wet, noisy brain?

To resolve these issues one must examine how well the possible quantum effects can survive in an environment that is potentially lethal to many of them. Careful analysis shows that one particular quantum effect, the 'quantum Zeno effect' can survive, and indeed can play an essential role in the causal relationship between a mind and its brain.

Of course, understanding *any* aspect of nature 'completely' may very well entail understanding all of nature completely. But this does not mean that understanding what physics alone can say about the mind–brain system completely entails understanding the psychologically described aspects completely. In fact, in the orthodox quantum description neither of the two kinds of aspects is, by itself, dynamically complete – rather, they complement each other. A specific problem is that within contemporary quantum theory the physical description does not by itself determine the occurrence or the character of certain *interventions* that are needed to complete the dynamics. In actual scientific practice the causal roots of these interventions are described in psychological terms, e.g., in terms of the intentions of experimenters. Thus, according to contemporary orthodox basic physical theory, but contrary to many claims made in the philosophy of mind, *the physical domain is not causally closed.* A causally open physical description of the mind–brain obviously cannot completely account for the mind–brain as a whole.

HA: In your articles you emphasize that your way to address the mind–matter problem does not go beyond what you like to call 'orthodox quantum theory'. However, quantum physics in its usual understanding excludes anything like mind, mental states, psyche, etc., even if issues of observation and measurement are discussed. Obviously, most experiments today are carried out in an entirely automatized way, so conscious human observers are not at all needed to register a measured outcome.

HPS: By 'orthodox quantum theory' I mean, specifically, versions of quantum theory (such as the original pragmatic Copenhagen interpretation, validated by actual scientific practice, and also von Neumann's extension of it) that explicitly recognize the fact that, *prior to the appearance of an experimental outcome*, a particular experiment needs to be set up. This 'setting up' *partitions* a continuum of quantum potentialities into a finite set of discrete possibilities. A simple example of such a partitioning is the placing of a detector of some particular size and shape in some particular location. The distinction between the firing and non-firing of this detector during some specified temporal interval then induces a bifurcation of a continuous space of potentialities into two subspaces, each correlated with a distinctive event, or lack thereof.

Von Neumann referred to this essential physical act of partitioning as 'process 1' and represented it in terms of projections onto different subspaces. Quantum theory depends upon the injection of such process 1 *interventions* into the dynamical evolution of the state of the system under study, which, except at the moments of these interventions, is controlled by the Schroedinger equation (which von Neumann called 'process 2'). An adequate theory of nature must accommodate physical process 1 actions even in situations in which no human agent seems to be involved. These interventions into the physical dynamics are perhaps the most radical innovation of quantum theory, vis-à-vis classical physics.

HA: If the formal structure of orthodox quantum theory remains unchanged in your approach, this can only mean that it also remains restricted to the material aspects of reality. This implies that, in order to include the mental domain, you have to invoke truly substantial additions to your framework of thinking, which are outside the realm of established physics. For this purpose you must have an ontology which (i) is consistent with our knowledge of (quantum) physics; (ii) allows a plausible incorporation of the mental, and (iii) provides ideas about how the two are related to each other – quite a program! How would you briefly sketch such an ontology?

HPS: In the first place, the structure of orthodox quantum theory allows us to make statistical predictions about correlations between initially known experimental conditions and the knowledge gleaned from their experienced outcomes. In Bohr's words (Bohr 1963, p. 60): "Strictly speaking, the mathematical formalism of quantum mechanics and electrodynamics merely offers rules of calculation for deduction of expectations about observations obtained under well-defined experimental conditions specified by classical physical concepts." In this sense, quantum theory concerns *directly* (i) the creation and experiencing of "well defined conditions specified by classical physical concepts"; (ii) the experiencing of outcomes of these actions; and (iii) certain predictions concerning relations among these two kinds of experiences. An adequate conceptual framework must provide an understanding of our role in the creation of conditions that will allow us to make quantum predictions pertaining to our resulting experiences.

In short, already the orthodox version of quantum mechanics, unlike classical mechanics, is not about a physical world detached from experiences; detached from minds. It is about predictions of relationships – entailed by a particular theoretical structure – between certain specified kinds of experiences.

The natural *ontology* for quantum theory, and most particularly for relativistic quantum field theory, has close similarities to key aspects of Whitehead's *process ontology*. Both are built around psychophysical events and objective tendencies (Aristotelian 'potentia', according to Heisenberg) for these events to occur. On Whitehead's view, as expressed in his *Process and Reality* (Whitehead 1978), reality is constituted of 'actual occasions' or 'actual entities', each one of which is associated with a unique extended region in spacetime, distinct from and non-overlapping with all others. Actual occasions actualize what was antecedently merely potential, but both the potential and the actual are *real* in an ontological sense. A key feature of actual occasions is that they are conceived as 'becomings' rather than 'beings' – they are not substances such as Descartes' res extensa and res cogitans, or material and mental states: they are processes.

HA: So what you suggest is to start from the ontologically neutral Copenhagen interpretation and supplement it with an ontology that is different from all other ontological interpretations of quantum theory that we know of. It combines Heisenberg's ontology of potentia with Whitehead's process ontology. Let us first talk about Heisenberg's ideas, and how they go beyond the picture of a materially tangible reality.

HPS: In his *Physics and Philosophy*, Heisenberg (1958b, p. 50) asked: "What happens 'really' in an atomic event?" He referred to events as happenings: "Observation [...] selects of all possible events the one that has actually happened [...]. Therefore, the transition from 'possible' to 'actual' takes place during the act of observation" (Heisenberg 1958b, p. 54).

Heisenberg's ontology is about sudden events and about 'objective tendencies' for such events to happen. The natural ontological character of the 'physical' aspect of quantum theory, namely the part described in terms of a wave function or quantum state, is that of a 'potentia' or 'tendency' for an event to happen. Tendencies for events to happen are not substance-like: they are not static or persisting in time. When a detection event happens in one region, the objective tendency for such an event to occur elsewhere changes abruptly. Such behavior does not conform to the philosophical conception of a substance.

Thus, neither the event nor its tendency to happen are ontologically substantive or self-sufficient: they are intrinsically connected to one another. Descartes' identification of two different 'substances' in reality is neither helpful for nor concordant with quantum theory. However, the conception of two differently described *aspects* of reality accords with both the theoretical and the practical elements of quantum theory.

HA: Whitehead's ontology is particularly radical insofar as it considers *process* as primordial, not substance – substance as understood in a philosophical sense. This is in contradistinction to all established sciences and almost all mainstream philosophy. How do you see the chances to establish a process ontology in the sciences?

HPS: Heisenberg never fully reconciled his ontological ideas with the epistemological stance of the Copenhagen interpretation. Chapter 3 of *Physics and Philosophy* (Heisenberg 1958b) is clearly an effort to bring these two aspects together. But to bring them successfully together in a rationally coherent and intellectually satisfying scheme requires one to say something about how the particular event that actually occurs comes to be selected.

Heisenberg did not address this issue, but Whitehead's account aims to explain it. Whitehead's fundamental process is the process of combining the pre-existing psychologically and physically described aspects of reality together to form a new psychophysical actual entity, or actual occasion, that is identifiable as an actual event (à la Heisenberg), whose physical manifestation is represented by a von Neumann process 1 action. I am merely proposing that Heisenberg's incomplete ontology be completed by accepting what I regard as Whitehead's main ideas. The aim of this approach is to understand how the psychological and physical aspects of reality conspire to select the events that actually occur. It allows the basically anthropocentric features of the pragmatic epistemological Copenhagen interpretation to be embedded within the general framework of a non-anthropocentric world process.

HA: So introducing Whitehead not only brings in process; it also, at the same time, integrates the psychologically described and the physically described aspects of reality into an overall processual dynamics.

HPS: Yes. And getting now to your question about the possibility of infiltrating these ideas in science, I need to stress that the core idea that the events in our streams of consciousness are two-way causally linked to events in the physical world lies at the intuitive heart of our daily dealings with reality. A theory that breaks this link is highly counterintuitive, and is also difficult to really make sense of, either in everyday life or in scientific practice.

School children during the mechanical age were readily able to accept the idea that the solid appearance of a table was an illusion; that the table was 'actually' mostly empty space, with tiny particles whirling around inside. How much easier will it be for future scientists growing up in the age of information, computers and flashing pixels to accept the idea of a world made of events and of potentialities for these events to occur?

My point here is that our most profound and deeply held intuition is not about the nature of the external physical world. It is rather that our human thoughts and efforts can make a difference in the behavior of our bodies. Our entire lives are based squarely on this perpetually re-validated intuition, as opposed to the proclamation of some philosophers, that our direct awareness of the physical efficacy of our thoughts is an illusion. The Heisenberg/Whitehead quantum ontology is thus concordant with both our most basic intuitions and with actual scientific practice. For this reason, I don't see why it should be difficult to shift science over to this improved way of conceptualizing nature and our role in nature.

HA: Whitehead treats matter and mind in terms of physical and mental poles of an actual occasion. This has the flavor of a dual-aspect approach, for which a number of other examples exist, such as Pauli's, Bohm's, Chalmers', or Velmans'. How do they differ from Whitehead's thinking, and from your own?

HPS: Pauli, in his discussion with Bohr about the notion of a 'detached observer', emphasized that the questions we pose cause nature some 'trouble'. The actions that instantiate these questions are the logically needed process 1 partitionings described by von Neumann.

My work carries forward Pauli's emphasis on this crucial point, but I remain so far uninfected by his speculations about archetypes and the like. Bohm's approach to consciousness brings in an infinite tower of explicate and implicate orders, each one 'in-forming' the one below and 'in-formed' by the one above. This picture is altogether different from the much more concrete Whiteadian quantum ontology. Chalmers appears to be moving in the right direction, but I believe he lacks a sufficiently firm grasp of quantum theory to be able to develop his approach in a way that I think would be fruitful. Velmans proposes an "ontological monism combined with an epistemological dualism" in which the quantum-induced failure of causal closure at the microphysical level is compensated for by a causal closure at the neurophysiological level. However, our conscious experiences are ontological realities in their own right, not mere epistemological bits of knowledge. So the claim of ontological monism seems unnatural, and the possibility of uncontrolled microscopic fluctuations exploding into uncontrolled neurophysiological fluctuations makes problematic the claim of dynamical completeness at the neurophysiological level.

But why go that route at all when quantum theory offers the possibility of bona fide straightforward real influences of conscious efforts upon physical brains, and consequently upon bodily behavior, without any demand of a causal closure of the physical at any level? Why hang onto one of the most controversial aspects of a materialist worldview, namely the notion that the causal efficacy of our conscious efforts is an illusion, when orthodox quantum theory seems to say just the opposite, and moreover provides the technical means for implementing the causal efficacy of our efforts?

HA: What about panpsychism, a key feature of both dual-aspect types of approaches and Whitehead's ontology? At which point in biological evolution do you think that the psychological aspect, the mental pole of actual occasions, becomes manifest? Or does it go all the way down to elementary particles?

HPS: Reduction events cannot act microscopically on individual particles. That would destroy the oft-observed interference effects. So we do not have end-to-end 'panpsychism'. Indeed, von Neumann's analysis of the measurement problem shows that it is nearly impossible to establish, below the level of human involvement, any failure of the unitary law (process 2) of purely physically determined evolution. The need for actual occasions even at the human level derives only from the philosophical commitment to accept as the foundation of objective science the outcomes of experiments "regarding which we are able to communicate to others what we have done and what we have learned" (Bohr 1963, p. 3). At present, we lack the empirical evidence needed to specify, on objective scientific grounds, the details of the embedding non-anthropocentric ontology which Whitehead's ideas demand. But we are certainly not yet at the end of science.

HA: As to the physical pole of Whitehead's actual occasions, you suggest a drastic reinterpretation, or better a major extension, of von Neumann's account of quantum measurement (von Neumann 1955/1932, Chaps. 5 and 6). While von Neumann discussed the physical aspects of measurement only, you refer to Bohr's and Heisenberg's distinction of (i) choices made by an observer (or experimenter) in terms of questions that are posed to nature and (ii) choices made by nature in order to answer those questions. The second aspect clearly refers to physics and places us in the role of detached observers, i.e., as 'impotent witnesses'. However, the first aspect introduces intentional actions by conscious human beings, at least if controlled experiments are discussed. As such, it escapes a purely physical discussion and points to the causal gap that you indicated above.

HPS: Von Neumann, the mathematician, described the purely physical aspect of the probing action, whereas Bohr, as physicist–philosopher, described the enveloping conceptual structure needed to tie the mathematical formalism to the activities and the knowledge of human beings. Bohr's pragmatic epistemology rationally accommodates the process 1 partitioning that is not understandable from within the causal framework provided by the mathematical formalism alone. This deficiency in the purely physical description is the causal gap. Bohr's pragmatic epistemology, eschewing ontological commitments, fills this gap by referring to the free choices of human beings. But Whiteheadian quantum ontology accepts *in reality* what Bohr accepts only pragmatically, namely the idea that our conscious intentions cause, at least in part, our intentional actions. This can be achieved by regarding the quantum reduction events to be the physical manifestations of the termination of a psychophysical process. Bohr's free choices are the psychological correlate of such a process 1 action, and, conversely, von Neumann's process 1 actions are the physical correlates of these conscious choices.

The physical and psychological aspects of reality are thus tied together in the notion of a quantum event.

Within orthodox thinking, the physical process 1 action results from, as von Neumann's words emphasize, an *intervention* from outside the physically described domain. This process has, according to contemporary quantum theory, no sufficient causal roots in the physical alone. The experimenter's 'free choice' *participates* in the selection of the needed partition that physical processes alone are unable to achieve. It is then the job of a satisfactory ontology to place these anthropocentric elements of human effortful attention within a broader non-anthropocentric conception of reality.

Ontological uniformity requires, plausibly, every such quantum event to have *some* experiential or felt component. But it does not require every actual occasion to have the full richness of a fully developed 'high-grade' human experience. The richness of the experience would naturally be expected to be correlated with the complexity of the physical system upon which von Neumann's process 1 acts.

HA: The correlation between physical state reduction (via projection) and mental subjective experience is posited as an assumption in your ontology, but it certainly does not follow from quantum theory! It is very much analogous to von Neumann's assumption of a psychophysical parallelism of brain and mind. Although von Neumann sometimes alludes to the mind ('abstract ego'), he actually refers to the brain in his discussion of quantum measurement.

HPS: Von Neumann focused on the mathematics, and avoided getting overly enmeshed in philosophical issues of interpreting quantum theory. But Heisenberg, speaking from the pragmatic epistemological perspective, said: "The observation itself changes the probability function discontinuously; it selects of all possible events the actual one that has taken place" (Heisenberg 1958b, p. 54). Thus, Heisenberg tied the mathematically described reduction events to the process of 'observation'.

In order to have a useful scientific theory one needs to link the mathematics to the perceptual aspects of our experience. The mathematical structure of quantum theory is such that the classical materialist accounts of the physical aspects of nature simply do not work. To achieve a conceptualization that ties the new mathematics to actual empirical scientific practice, in a rationally coherent and practically useful way, the founders of quantum theory switched to a conceptualization of the physical world based upon empirical events, such as the click of a Geiger counter, and upon potentialities for such events to occur. The mathematics thereby becomes linked to empirical phenomena within the theory itself.

HA: The notion of an interaction between mind and matter, as in your recent paper (Stapp 2005) on 'interactive dualism', may be somewhat misleading. It seems to me that things are much more subtle than a straightforward interaction between the mental and the physical (which one might naively interpret as basically similar to a collision of billiard balls). The proposal by Eccles, whose physical features were worked out by Beck, has this overly simplistic appeal because some 'mental force' is assumed to act directly on synaptic, i.e. material, transport processes. Your picture is definitely much more subtle: the requirement that physical and mental outcomes of an actual occasion must match, i.e., be correlated, acts as a constraint on the way in which these outcomes are formed within the actual occasion. So the notion of an interaction should be replaced by the notion of a constraint set by mind–matter correlations.

HPS: It would indeed be misleading to understand the 'action of mind upon brain' directly via a 'force'. The effect is associated with a modulation of the frequency of certain process 1 actions that act directly upon large-scale (brain-sized) patterns of neurological activity. This modulation of frequencies is achieved, strictly within the pragmatic framework (that is, without any of Whitehead's ontological superstructure) by exploiting certain human 'free choices' that are allowed within that pragmatic framework. This language suggests that the conscious act is the cause, and the correlated physical process 1 action is the effect. This interpretation ties the theory most naturally and directly to actual scientific practice. In actual practice the experimenter chooses on the basis of reasons and goals which of the experimental options will be pursued, within the array of possibilities that the structure of the physical theory provides. Bohr (1958, p. 73) spoke, accordingly, of "the free choice of experimental arrangement for which the mathematical structure of the quantum mechanical formalism offers the appropriate latitude". We are dealing here with the sophisticated way in which mental intention influences quantum processes in the brain. *Ideas* do not simply push classically conceived particles around!

HA: A major point in your ontological framework is that physical state reduction and mental subjective experience jointly constitute the transition from the continuous and the potential to the discrete and the actual. Another significant issue is the contrast between instantaneous projections, which von Neumann introduced as an idealization that he characterized as 'not enjoyable', versus an objective dynamical process of measurement that takes time, as advocated by a number of physicists. For instance, Penrose strongly argues that way in his speculations about mind and matter. Of course, this would require an individual rather than a statistical description of quantum measurement, of which no broadly accepted version is available so far.

HPS: The mathematical neatness of instantaneous (along a spacelike surface) reduction makes it the better option, technically and mathematically, and I see no reason to complicate the dynamics by smearing out in time the reduction events. Indeed, to do so would confuse everything, since the smearing would not be strictly confined, and hence process 2 would never hold rigorously.

The fact that we experience process as involving duration is adequately explained by James' 'marching band' metaphor. Each instantaneous 'snap shot' corresponding to a single experience would catch the components of brain activity correlated with the various stages from just beginning to be experienced, to full blown vivid consciousness, to fading out. This structure creates the impression that the experience has duration, although it is really instantaneous – or confined to a spacelike surface, when mapped into real spacetime (Stapp 1993/2004, Sect. 6.6).

HA: For details of what happens at the mental pole of an actual occasion,the notions of attention and intention according to William James in combination with your concept of a 'template for action' figure prominently in your work, e.g., in Stapp (1999) and in Schwartz, Stapp, and Beauregard (2005). Could you outline how these terms are related to one another?

HPS: A *template for action* is defined to be a macroscopic (extending over a large portion of the brain) pattern of neurological activity that, if held in place for a sufficiently long period, will tend to produce a brain activity that will tend to produce an intended experienced feedback. This pattern of brain activity is the neural correlate (specified by a process 1 action) of a conscious effort to act in an intended way. William James (1892, p. 227) says that "no object can catch our attention except by the neural machinery. But the amount of attention that an object receives after it has caught our attention is another matter. It often takes effort to keep mind upon it. We feel we can make more or less of the effort as we choose. [...] This feeling [...] will deepen and prolong the stay in consciousness of innumerable ideas which else would fade away more quickly."

Effort is a particular feature of consciousness that we feel we can control, and that has the effect of intensifying experience. Hence it is reasonable to suppose that increasing effort increases the rate at which conscious events are occurring. If the rate becomes sufficiently great then the quantum Zeno effect will, according to the quantum laws, kick in, and the repetitious interventions of the probing actions will tend to hold in place the template for action. That effect will, in turn, tend to make the intended action occur. By virtue of this dynamically explained causal effect of willful conscious effort upon brain activity, trial-and-error learning should hone the correlation between the consciously experienced intention and an associated template for action that produces, via the physical laws, the intended feedback. This *explains* dynamically the capacity of an effortful intention to bring about its intended consequence.

HA: From a psychological point of view, one might distinguish a series of steps: from a mental state with a particular intensity of attention to the shift of that attention and finally to an intention to make a decision, which is correlated with a neural template for action. This template precedes the action – it is not already the action itself. Are there empirically accessible psychological observables for these different steps?

HPS: Actions include brain actions that control or guide other brain actions. The theory says that each of the different experienced stages should occur in conjunction with a different template for action. For instance, the actualization of one early template could tend to set in motion a multi-component sequence leading from neural activity somewhere in the cortex to activity in the motor cortex to muscular activity.

HA: Concerning the neural correlates of such psychological states and observables, we need the notion of a neural assembly. If you assume that such neural assemblies are subject to a quantum Zeno effect, this requires that they be in an unstable state, such as a quantum superposition, or an entangled state. How do you think this condition can be realized for an assembly of thousands of neurons?

HPS: Environmental decoherence effects will reduce the entire brain state in question (represented by a reduced density matrix) to a statistical mixture of states each of which is essentially a slightly smeared out classically conceived possible state of the entire brain. This decomposition of the state of the brain into a mixture of almost-classical states is very useful in connection with this theory. It allows neuroscientists to quite accurately conceive of the brain as a collection of almost-classical possibilities that continually diffuse into more diversified collections, but that are occasionally trimmed back, in association with a conscious experience, to the subcollection compatible with that experience. These processes all involve, or can involve, assemblies of thousands or millions of neurons.

HA: What do you mean by "slightly smeared out" and "almost-classical"? If you have some remaining quantum features in the brain state – which you need for the quantum Zeno effect to act – you must assume that the brain state was a quantum state to begin with. How is such a state constituted, or prepared? Or do you assume that *every* system is fundamentally a quantum system which, under the influence of its environment, decoheres more or less rapidly into classical subsystems?

HPS: By "slightly smeared out" and "almost-classical" I mean what you would get from a classically conceived state if you replaced each point particle by a very tiny continuous cloud of possibilities. Each physical system – including a brain or a template for action – inherits quantum features from the quantum state of the universe as a whole. In the case of a brain, decoherence mechanisms are acting strongly at all times, and they never allow its state to be anything other than a mixture of almost-classical (i.e., slightly smeared out) states. Hence the classical intuitions of neuroscience about the brain are generally valid, except for two things. Firstly, at almost every instant the cloud of possibilities is growing and diffusing into a wider set of possibilities which, however, every once in a while (at a reduction event) gets reduced to a subset. Secondly, the diffusing action can be curtailed by the quantum Zeno effect which arises from the small, but nonzero, quantum smearing of each one of the almost-classical components.

In this way, the brain is described strictly quantum mechanically, yet it can be understood to be very similar to a classical statistical ensemble. Importantly, the relevance of the quantum aspects for consciousness is not due to some macroscopic quantum superposition effect, which would be extremely hard to realize. The pertinent non-classical feature is the occasional occurrence of a sudden reduction of the ensemble to a sub-ensemble that is compatible with the content of a co-occurring conscious experience.

The occurrences of such reductions are logically *possible* because the state of the brain represents not an evolving material substance but rather an evolving set of potentialities for a psychophysical event to occur. The occurrences of such reductions are logically *necessary* because the expanding ensemble of almost classical states is a continuous structure that must be decomposed into a collection of discrete alternatives, each associated with a distinct kind of experience. It is only by means of this partitioning that the theory is tied securely to human experiences, and to the empirically validated rules of quantum theory. The smear of almost-classical possibilities must be partitioned, prior to each experience, into a specified collection of components at least one of which corresponds to a distinctive experience, or lack thereof.

HA: As you said before, brain states or templates for action cannot be Zeno-stabilized simply by the direct action of something like a mental force – this would lead to the same basic problem that Eccles has with his proposal for a direct mental influence on synaptic processes. So what do you concretely assume at the neural level that is capable of exerting a quantum Zeno effect upon the template for action?

HPS: As an example, let us suppose that the occurring process 1 action partitions the state of the brain into two parts. One of them, the 'Yes' part, is the neural correlate of the mental intent to, say, 'raise the arm'. This neural correlate is a template for action. The immediately felt psychological effect of an increased intentional effort is an intensified experience of the projected intended feedback. These *projected* experiences are constructed from the memories of earlier experiences, as discussed in Stapp (1993, Sect. 6).

Now the *timings* of the process 1 actions are an aspect of the 'free choice' on the part of the human observer. It is therefore plausible to conjecture that the effort-induced increase in the intensity of the projected intended experience is *caused* by an increase in the observer-controlled rate at which the associated process 1 actions are occurring. If the essentially identical process 1 actions occur in sufficiently rapid succession, then the associated neural correlate (i.e., the template for action) will be held in place by the quantum Zeno effect. The resulting *persistent* neural pattern of activity will then tend to cause the intended action to occur. The *effect* of the effort-induced increase in the rate of the process 1 probing actions is thus to hold in place the entire macroscopic template for action. The dynamical effect, via the neural machinery, of this holding in place is the likely occurrence of the intended action.

This scenario differs in two important ways from the proposal by Beck and Eccles. First, the action does not take place at the synaptic, i.e., *microscopic*, level: the effect is directly upon the entire template for action, specified by von Neumann's process 1 action. And, in answer to your question about 'mental force', there is no action of any forces, mental or otherwise, upon the parts of a material substrate: no pushing around of the atoms in a way that produces, in some totally miraculous and unaccountable way, the action that the person has in mind. No! The effect of the effort is on an entire macroscopic neural pattern of brain activity. This pattern has been singled out by von Neumann's process 1 action and is held in place by the quantum Zeno effect. By coupling von Neumann's dynamical rules to learning, one can rationally account for the observed – and essential for human life and survival – correspondence between experienced intent and experienced feedback.

HA: After all, this amounts to an overall theoretical picture that offers a strong sense of formal and conceptual coherence and is intuitively appealing in a number of respects, but also confronts us with a remarkable degree of complexity. What do you think: Is there any chance that empirical work can confirm or falsify particular features of your approach?

HPS: First of all, it is evidently forever impossible to falsify, by empirical data alone, the opposing blatant assertion that the apparent causal efficacy of our conscious efforts is an illusion. It is impossible to disprove empirically the physicalist contention that our conscious experiences are merely causally irrelevant pyrotechnics that *seem to be* influencing the course of bodily events, but are, in reality, merely impotent by-products of causally self-sufficient neural activities. But what rational argument could adequately justify such an outrageous and completely unsupported claim? Like solipsism, such a claim cannot be empirically falsified, but only rejected on the grounds of its lack of reasonableness and utility.

During the nineteenth century, before the precepts of classical physics had been shown to be *fundamentally* false, scientists and philosophers had some sound reasons to conjecture that the physical aspects of reality were causally closed. However, even then this led to an unreasonableness noted by William James (1890, p. 138): consciousness seems to be "an organ, superadded to the other organs which maintain the animal in its struggle for existence; and the presumption of course is that it helps him in some way in this struggle, just as they do. But it cannot help him without being in some way efficacious and influencing the course of his bodily history." James went on to examine the circumstances under which consciousness appears, and ended up saying: "The conclusion that it is useful is, after all this, quite justifiable. But if it is useful it must be so through its causal efficaciousness, and the automaton-theory must succumb to common-sense" (James 1890, p. 144).

That was James's conclusion even at a time when classical physical theory seemed irrefutable, and the thesis of brains as mechanical automata was rationally supported by physics-based legs. Today, however, classical physics has been superseded by a theory with causal gaps that *need to be* filled in some way, and that *can be* filled by allowing our efforts to do what they seem to be doing. Embedded in an adequate ontology, quantum theory has the technical capacity to explain how a person's conscious efforts can influence his or her bodily actions. Consequently, there is now no rational reason whatsoever to reject William James's persuasive argument.

Beyond these philosophical considerations one can reasonably claim that the entire body of neuropsychological experimentation is confirmatory of this theory. All the data, to the extent that they are precise enough to say anything about the relationship between mind and brain, are in line with this theory. A large number of particular empirical findings in neuropsychology and in the psychology of attention are discussed in Schwartz, Stapp, and Beauregard (2005).

HA: The current support for this novel picture, especially as far as cognitive neuroscience is concerned, is merely qualitative though.

HPS: Well, there are plenty of ways to falsify the quantum model. It demands close connections between the psychological and the physical aspects of psychophysical events. This includes, in particular, the putative attention-induced quantum Zeno effect of holding in place the templates for intentional action. But there is evidently no way to counter the claim that *whatever* connections are empirically found are exactly what the *it's-all-an-illusion* proponents could claim that their theory allows. For that position has no theoretical foundation in established physics, and hence no basis upon which to falsify it.

Many scientists and philosophers have forced themselves to accept the rationally unsatisfactory and unsupported physicalist position in the mistaken belief that this is what basic physical theory demands. But the converse is true: contemporary physical theory demands certain interventions into the physical! The associated causal gap in a purely physically determined causation provides a natural opening to an interactive but non-Cartesian dualism.

HA: Since your approach does emphatically refer to attention, intention and, if I may use this term, 'free will', it must have ethical implications. Would you say that proper ethical behavior can be facilitated or even entailed by reflecting and realizing the ontological conditions of a given situation? Or, conversely, is that ethical misbehavior a consequence of lacking insight of appropriate ontological conditions? Might a processually conceived quantum theory, comprehending both psychological and physical aspects of nature, provide insights that could underpin a science-based rational ethical theory?

HPS: Behavior, insofar as it concerns ethics, is guided by conscious reflection and evaluation. It is not mere unreflective response. The output of such reflections and evaluations depends, of course, on the input, and the core of the effective input is the individual's self-image in relation to his or her conception of the situation in which he or she is embedded. One's weighting of the welfare of the whole, and one's sensitivity to the feelings of others will surely be enhanced when the individual sees his or her own judgments and efforts as causally effective – hence important – inputs into a cooperative effort to develop the vast yet-to-be-fixed potentialities of a quantum universe that, as

Bohr emphasized, can be properly conceived only as an intricately interconnected whole.

Such a comprehension of self stands in strong contrast to an image of the self as a cog in a pre-ordained mechanical universe – a cog that thinks of his or her strenuous efforts to choose rightly as making no actual difference whatever in the course of physical events. Such a degradation in self-image will undoubtedly be correlated with a debasement in behavior. Conversely, what you call ethical misbehavior would surely be diminished by a shift in self-image from mechanical cog to quantum player.

HA: If this is extended beyond individual human beings, it must also have implications for human societies and their ways to interact with each other.

HPS: The proposition, foisted upon us by a materialism based on classical physics – that we human beings are essentially mechanical automata, with every least action and thought fixed from the birth of the universe by microscopic clockwork-like mechanisms – has created enormous difficulties for ethical theory. These difficulties lie like the plague on Western culture, robbing its citizens of any rational basis for self-esteem or self-respect, or esteem or respect for others. Quantum physics, joined to a natural embedding ontology, brings our human minds squarely into the dynamical workings of nature. With our physically efficacious minds now integrated into the unfolding of uncharted and yet-to-be-plumbed potentialities of an intricately interconnected whole, the responsibility that accompanies the power to decide things on the basis of one's own thoughts, ideas, and judgments is laid upon us. This leads naturally and correctly to a concomitant elevation in the dignity of our persons and the meaningfulness of our lives. Ethical theory is thereby supplied with a rationally coherent foundation that an automaton account cannot match.

But beyond supplying a rational foundation for Western culture, the rooting of ethics in science, with its universal character and appeal, shifts values toward the ecumenical, and away from those aspects of religions that are hostile to, and preach violence against, followers of other faiths. Such a shift is sorely needed today.

15 Consciousness and the Anthropic Questions

By the anthropic questions I mean the following three queries:

1. Why are the laws of nature so well tuned to support the biological structures we find here on earth, including our own human bodies and brains?
2. Why, given the fact that the physically described structure of my body and brain has the form that it has, are certain activities of that physically described system accompanied by my stream of conscious experiences, which convey the pervasive impression that elements of this stream of experiences causally affect the way my body behaves?
3. Are idea-like qualities primordial? Or do they emerge from a world completely devoid of all mind-like qualities?

These questions may lack answers that human minds can comprehend, or that our scientific investigations can find firm evidence to support. Still, these questions are being asked, within scientific contexts, and a science-based world view may be incomplete without some rationally coherent responses to them.

It is evident to me that such basic questions must be addressed within the framework of *basic* physical theory, namely quantum theory, not within the conceptually alien theory obtained by making the classical approximation to it. For quantum theory is critically involved with the stability of matter that underlies our bodily existence, with the properties of organic molecules, such as DNA, that underlies life, and with the "element of wholeness, symbolized by the quantum of action, and completely foreign to the classical physical principles" that characterizes a wide range of empirical phenomena. The classical approximation rationally supports neither stability, life, nor the quantum wholeness aspect of our conscious observations

The original Copenhagen interpretation of quantum theory was designed to deal specifically with scientific practices in which experimenters first set up experiments of their choosing, and then test predictions of the theory. These predictions pertain to observed correlations between the empirically described initial conditions that these experimenters have set up, and the empirically described feedbacks that they then observe. This format carries over to the situations in everyday life pertaining to correlations between how we choose to act and the feedback we are then likely to experience, but does not carry over easily to issues such as the origin and development of the physical universe, of life, or of consciousness. Yet scientists ponder these further issues and do try to create theories that can account for the available scientifically acquired evidence concerning them.

The focus of this book has been primarily on the dynamics of human brain–minds *as they exist today*. However, the appearance of the interview by Atmanspacher in the Journal of Consciousness Studies prompted a flurry of questions posted on jcs-online. These questions and my replies to them lay the foundation for some responses to the anthropic questions. I shall retain here the order in which the questions were posed online, because the weaving back and forth between different threads creates in the end a richer tapestry that better conveys the whole.

[Jonathan Edwards] I fail to follow Stapp's suggestion that the 'free choice' of setting up an experiment is somehow outside the scope of a physical account. True, the events in a brain freely choosing are too complex to analyze, but that does not put them outside physics. This is no evidence that physics is not causally closed.

HPS: Edwards says that he fails "to follow my suggestion that the 'free choice' of a setting up of an experiment is somehow outside the scope of a physical account". What I claimed was that this choice is outside the scope of conventional or orthodox quantum theory. The conventional quantum theory that is used in actual scientific practice requires an intervention from outside the system of the atomic constituents that are described in the mathematical language of quantum physics. Bohr calls this intervention a 'free choice' on the part of the experimenter, and von Neumann calls it process 1. The *effects* of such interventions enter importantly into the dynamics of the quantum mechanically/mathematically described physical system, and into the structure of our subsequent conscious experiences. But that theory provides no explanation or causal description of how these causally effective choices come to be what they turn out to be. And the omission arises from more than just the fact that brain processes "are too complex to analyze". If one considers the experiment discussed by Einstein where a pen is drawing a line on a moving scroll, with a 'blip' being caused by the firing of a detector of a slow radioactive decay, then the deterministic Schroedinger equation, applied to the whole system, would yield a smear of times for the registered decay, and, likewise, a corresponding smear of the location of the blip. And if an observer is watching the device, and his body and brain and its environment (all of which are made up of atoms, molecules, and other physically describable constituents) are incorporated into the quantum mathematical description, with no intervention from outside, then the system described by the quantum state of the brain of the observer becomes a smeared out mixture of quasi-classical brain states that correspond to different possible times for the detection to occur.

Conventional quantum theory provides no purely physical description of how this smear of brain states gets reduced to one compatible with experience, which identifies a fairly well defined time of registration. Conventional theory, as defined either by actual scientific practice or by the words of the founders of the theory, has, therefore, a *causal gap* in the purely physical description, and this gap is not simply a matter of the quantum-mechanically described physical workings of the brain being "too complex to analyze". There is a matter of principle involved in understanding how a brain state in which the recognition of the blip is smeared out over hours turns into a brain state that corresponds to the registration occurring at some particular moment.

The point is that in conventional quantum theory the quantum mathematical description becomes a description merely of possibilities or potentialities, not, in general, of an evolving experiential reality itself. Yet this quantum state is precisely the quantum theoretical generalization of the classical-physics description of physical reality itself. So where did the 'physical reality' itself go? What is the rational basis of the claim that the physical description is causally closed when the classical physics description, from which the notion of the causal closure of the physical arose, dissolves into mere potentialities, and the only realities – as opposed to potentialities – that are to be found in the phenomenally validated conventional quantum theory are described in psychological rather than physical terms?

[Edwards] Stapp's quantum-Whiteheadian 'events' seem one minute to belong to observers as whole brains, at other times to local neural events and sometimes events outside brains, while baulking at panpsychism because 'observers' have to be large enough to be 'classical'.

HPS: Edwards complains that I associated the collapse 'events' sometimes with the whole brain of the observer, sometimes with neural events, and sometimes with events outside the brain. I do not associate the collapse events with events at the individual neural level: I discussed nerve terminal dynamics only to show that the classical approximation fails in principle, i.e., only to show that one must *at least in principle* treat the brain as a quantum system. The collapse events in conventional quantum physics are, in fact, *psychophysical*: each one has both a psychologically described aspect, corresponding to an increase in knowledge, and also an associated reduction of the (physically described) wave packet (quantum state) to one compatible with the gain in knowledge. This is how the theory works in actual scientific practice. This arrangement ties the psychological descriptions of the events, which specify gains in knowledge and commitments to actions, to the effects of such inputs on potentialities for future human experiences, which are the only realities – as opposed to potentialities – that enter into the empirically justified conventional quantum theoretical description.

I do not use 'classical' as a condition on the physical aspects of the psychophysical events. And these events always occur '*in* a brain', in the sense that its physical aspect is represented in terms of 'projection operators' acting on the physically described state of some physical system, which can be called, generically, a 'brain'. I am simply retaining and applying von Neumann's quantum dynamical rules and the mind–brain connections that they imply.

[Edwards] A further problem is that Stapp gives no neurobiological example of what sort of 'events' he is implicating.

HPS: Regarding the nature of the neurobiological 'events' that are associated with our human intentional choices, I note that these events are expected by most (or at least many) scientists working on this problem to be the coming into being of widespread synchronous cortical activities in the beta or gamma frequency range. These neural activities are good candidates for the 'templates for action' that are the neural (brain) correlates of our conscious intentions. This matter is currently under intense empirical and theoretical scrutiny.

[Edwards] Stapp seems to view the alternative as epiphenomenalism, but as far as I can see the real alternative is that consciousness fits into current physics as an aspect of a causal chain in a way yet to be understood which happens to have nothing to do with the collapse of a wave function belonging to some particle that has hit a measuring device somewhere in a physics lab – which always seemed a rather odd idea anyway to me.

HPS: But is the 'current physics' quantum physics or classical physics? If it is classical physics then since the concepts of "conscious thoughts, ideas, and feelings" do not enter into the causal chain described in classical physics, which in principle is dynamically deterministically complete, these experiential realities are not part of the classically described causal chain: they are properly and correctly called 'epiphenomenal' insofar as that classical physics description of the causal dynamics holds. If by 'current physics' one means a physics described in physics text books and taught in physics courses at our universities, then, if it is not classical physics, 'current physics' presumably means quantum physics. If one is talking about mind–brain connections then one is talking about the physics of the brain. The pertinent events are events in the brain, not, primarily, events of particles hitting measuring devices in the lab, although such an event may lead by a physical causal chain to an effect in the brain, including the physical event in the brain that corresponds to the recognition that the detector reacted. Von Neumann followed the causal chain from the particle-detector event to the brain event that corresponds to conscious recognition. That *latter* event is the quantum event of primary interest, if one is interested in the mind–brain connection.

In current contemporary orthodox quantum theory the collapse events are closely connected to human experience: it is precisely this close connection that makes the theory practically useful. In view of the psychophysical direction that the hugely successful advance from classical physics to quantum physics took, it is reasonable and rational to retain the theoretical interlocking of physically described and psychologically described aspects that constitute a core radically new element of the improved theory.

Science must deal, of course, with relationships between *descriptions*. The dynamical laws of classical physics specify no direct dynamical connection between these two kinds of descriptions upon which science is based. Quantum theory, on the other hand, incorporates explicitly and nontrivially into its dynamical laws connections between these two kinds of description.

[Edwards] I find it hard to identify exactly what Stapp is proposing but the impression that comes across is that it is a form of dualism far more extreme than Descartes. Descartes tried to fit the mind into naturalism. In contrast, Stapp proposes a Deus ex Machina ghost in the machine; overtly supernatural in a way that is certainly pre-Darwinian and probably pre-enlightenment – i.e., medieval. This ghost is a religious object that has to be taken on faith because the very theory that is used to propose it denies the possibility of testing its existence.

HPS: Edwards suggests that bringing conscious realities into the description of nature in the way I describe (which is exactly the way prescribed by conventional quantum theory) is 'overtly supernatural'. But conventional quantum theory is highly naturalistic, and is closely tied to empirical phenomena. There is nothing supernatural about the conscious realities that populate our streams of conscious experiences, and nothing non-naturalistic about bringing these realities into physical theory in the mathematically specified, testable, and highly tested way specified by conventional quantum theory. That theory, as it stands today, is ontologically and dynamically incomplete, because it does not explain or describe how our specific choices about how we act come to be what they turn out to be. Recognizing this incompleteness is an act far different from postulating a 'ghost' that "is a religious object that has to be taken on faith". Our actual conscious intentional acts are not ghosts. They have theoretically explained and empirically measured consequences. Emphasizing the fact that these realities enter conventional quantum theory as causally effective inputs whose causal origins are "yet to be understood" is not an act of religious faith: it is the assertion of a basic fact about the current state of physical theory.

[Edwards] The idea that suggesting that QM is supported by ghostly puppeteers 'choosing' what will happen in our bodies can help us to be ethical seems as crazy as creationism.

HPS: My suggestion is that replacing the classical-physics conception of oneself (as a being that is causally equivalent to a mindless mechanical automaton stalking through a mindless clockwork universe) by the quantum conception of oneself (namely as an integral aspect of nature's non-local process of creation that allows components of one's stream of consciousness, such as reasons and values, to influence the activities of one's brain and body) provides a rational foundation for the notions of *responsibility* and *belonging to a community* based on trans-cultural contemporary science rather than on culture-dependent and often antagonistic religious faiths. Causally efficacious mind is a prerequisite of ethical theory, and quantum theory allows it to be supplied by science, rather than by a religious faith or doctrine that contradicts science, *insofar as science is identified with classical physics*. With respect to the notions of the efficacy of our conscious efforts, of personal responsibility, and of the concept of community, contemporary science has important commonalities with the major religions.

[Joseph Polanik] Dr. Stapp, You've described your theory as a quantum interactive dualism. I am wondering whether you are proposing a dualism as Chalmers counts or a dualism as Descartes is commonly thought to have counted. Chalmers, as I understand it, assumes that there is only one 'stuff' matter/energy; but, that this 'stuff' has two sets of properties – the physical properties familiar to scientists and the experiential properties associated with phenomenal awareness. Descartes, on the other hand, had two fundamental substances matter/energy and mind stuff. He is usually thought to have two sets of properties; but, some argue that he had three sets of properties: physical properties, mental properties and the experiential properties associated with sensory awareness that came about because of the union of body and mind/soul. In any event, there is no reasonable way to define 'dualism' so that Chalmers and Descartes are in the same camp. Thus, the question that arises is: Is your dualism a Chalmers-style dualism, a Descartes-style dualism or something else?

HPS: I start from the structure of conventional quantum mechanics; the quantum mechanics used by physicists in their scientific practice. It uses two kinds of *descriptions*. One kind of description is used to "comunicate to others what we have done and what we have learnt" (Bohr 1962, p. 3). It is basically a description of the thoughts, ideas, and feelings that populate our streams of conscious experiences. It is a description of psychological qualities. The other kind of description is the quantum mathematical description in terms of mathematically characterized properties assigned to spacetime points. These physical descriptions are the mathematical elements upon which theoretical physics is based.

Each of these two descriptions might be said to be describing an aspect of nature that possesses a certain persisting 'essence', psychological or physical, although neither aspect is a 'substance' in the normal/usual everyday sense of the word. The physical descriptions specify 'potentialities' for psychophysical events to occur. These events are the only *actual* things represented in the conceptual structure. Each such *event* is supposed to have both a psychologically described aspect, and a physically described aspect. The latter is expressed as a 'reduction' of the prior set of potentialities to a new set compatible with the gain in knowledge described in the psychologically described aspect.

Having described the structure of conventional practically validated physical theory, I leave to you the task of applying the appropriate labels from the philosophical literature.

The key point is that the theory deals with the *descriptions* themselves, not with what the descriptions are descriptions of. So it is sufficient to say that nature is a process that has aspects that we can and do describe in two different ways, psychological and physical. Quantum mechanics specifies a way of linking the descriptions that we use "to communicate to others what we have done and what we have learnt" (Bohr 1963, p. 3) to a certain theoretical description of mathematical properties assigned to spacetime points.

Chris Nunn raises a pertinent point: Can the proposed attention-controlled rate of process 1 probings be rapid enough to activate a quantum Zeno effect that is adequate to the task of holding the template of action in place? Chris suggests that the rate of probings ought not be more than about 100 hertz, which could be far less than what would be needed to hold in place a macroscopic template for action.

But, of course, we do not feel or experience the separate probings directly: we feel/experience only the conscious effort and the correlated phenomenal events.

A typical classically describable and observable electromagnetic field, say of fixed frequency and energy, has two very different frequencies associated with it: the classical frequency that is directly observed, and the quantum frequency determined by its energy. The 100 hertz mentioned by Chris is the classical frequency, but there is also the typically much greater quantum frequency.

In the model I am describing the directly experienced aspects, the intent and the experienced correlated feedback, have a time scale of at least tens of milliseconds: the time scale of the classical frequencies. In a recent application of von Neumann's theory to empirical data pertaining to binocular rivalry Efstratios Manuosakis has found a sensitive good fit to data that depends heavily upon a quantum Zeno effect, and that uses 10 milliseconds for the time between probing actions. In the model the pertinent quantum states all have the same frequency and are almost stable. These conditions allow the probing frequencies to be $\sim$ 100 hertz.

Mondor questions, in connection with the phrase 'free choice', the concept of 'free'. First, regarding my use of the word 'free' in 'free choice', I have repeatedly emphasized that this word refers here specifically to the fact the causes of these choices are not specified by conventional (Copenhagen or von Neumann) quantum mechanics. The word is not meant to suggest that these choices have no causes at all.

I believe that nothing happens without a sufficient reason of some sort, and my basic endeavor, in fact, is to try to achieve some deeper understanding the nature of these reasons.

Polanik hits the mark when he emphasizes that the physically described brain (governed by von Neumann process 2) is not self-collapsing, and hence that something beyond quantum mechanically described physical brain, evolving in accordance with the quantum mechanical counterpart of the classical laws of motion, is involved in the selection, from the smeared out mixture of potentialities generated by this mechanical process, the particular experience that actually occurs

Von Neumann's discussion of the movable boundary between the part of the world described in physical terms and the part described in psychological terms stresses 'psychophysical parallelism': the fact that certain systems have aspects that are described in the mathematical language of QM, and also aspects described in the language of communication among observing and acting agents. In classical physics the part described in physical terms, when expanded to include the entire physical universe, is causally closed – it is deterministically complete. But the core feature of quantum theory is that this causal closure does not occur within this more accurate theory. The whole is not fully causally described in either one of these two languages alone.

When Lofting says that we must "step out of the QM box and into the *general* box of how our neurology processes data" he seems to be assuming, unjustifiably, that the causal structure is describable in physical terms alone. The structure of quantum theory opens the door to the possibility that all causes and reasons need not be purely mechanical. Thoughts and intentions are themselves actual realities, and as such they ought to be able have, in their own right, real actual consequences. Quantum theory allows this, and in actual scientific practice demands it.

Feynman, mentioned by Edwards, asserted that he did not understand quantum mechanics, and doubted that anyone else did. The problem is basically the mismatch between the known basic purely physically described laws and our conscious experiences. QM tells us that when we try to descend to the microscopic roots of the 'physical substrate' the physically described properties dissolve into potentialities for the occurrences of experiences. The suggestion by Edwards that Bohr, Heisenberg, Pauli, Wigner, and von Neumann were introducing 'fairies' into basic physical theory, by introducing our experiences importantly into basic dynamics, is unhelpful: the entry of causally efficacious consciousness coherently into physics ought not be treated lightly. Nor does the emphasis on the fact that "we do not yet know how our choices of how we shall act come into being" entail or suggest that those causes, whatever they are, are unnatural.

Nunn correctly observes that the theory entails a person's capacity to choose to sustain a desired macroscopic brain activity without that person's knowing the physically described details of what his choice is actually doing.

Trial and error learning allows the person to correlate his mental effort to experienced feedback without his knowing how that conscious effort produces that conscious feedback. The mechanism that I am proposing merely requires that when a conscious event occurs that features a conscious intention to act in some way, coupled with an 'evaluation-based' feeling of effort, then there will be a tendency for the *action* associated with that feeling, *whatever that action is or was*, to occur repetitively with a frequency controlled by the intensity of the feeling of effort. This allows the agent to choose to sustain positively valued actions without knowing the actual physical form of the collapse events that his or her efforts are causing. This theory accommodates nicely and naturally the experience, for example, of learning to use a prosthetic limb, by activating through effortful trial and error learning a conscious-effort/conscious-feedback loop never before used either by the individual or by any of his or her ancestors.

The claim here is not that the physically described learning could not be explained mechanically. One could presumably devise some classically conceived process that could reproduce the main features of the observed behavior. But in that mechanical description consciousness plays no role. However, the point is that there is now *also* an alternative possible science-based explanation in which conscious effort is causally efficacious in the way that it feels to us that it is.

So which of the two putative explanations is scientifically better? One must be careful here. A snap judgment might be that the classical-physics-based explanation is better, by virtue of Occam's razor, because it does not bring in consciousness, and hence is more economical.

But there are two overriding considerations. The first is that biological systems in general, and brains in particular, are inherently quantum mechanical, featuring unstable elements and feedback mechanisms that tend to magnify explosively the quantum uncertainties of the underlying ionic processes. This destroys, from a scientific perspective, the microscopic deterministic underpinnings of the classical-physics model of this system. The classical model may be simpler, but, fundamentally, it cannot be physically correct. Certain simple features of atoms can be explained *more economically* by using classical physics, but that does not make the classical description of the atom correct.

The second, and even more decisive, consideration is that applying Occam's razor in this way is too drastic, because it eliminates not simply a removable theoretical construct but the known actual realities – our conscious thoughts – which are, furthermore, the principle subjects of interest. What we want to know, here, are how our thoughts are related to the activities of our brains. Hence reverting to classical physics is disastrous because it cuts the linkage between brain activity and consciousness that the more accurate quantum theory describes.

Nunn asserts that: "It's not obvious that this [quantum approach] provides any better grounding for a naturalistic concept of free will than classical mechanistic accounts of brain function."

But the 'free will' problem is to explain the connection between 'willed' physical actions and the conscious efforts that *seem* to be causing them. How, in the classical approach, does one explain how the conscious thoughts come into being at all, and in just the way that makes them appear to be doing just what, according to quantum mechanics, they actually are doing? And how can they, during the evolution of our species, *evolve* in just the right way as to be always in accord with what the brain is causing the body to do if there could be no adverse physical consequences of the thoughts themselves going completely haywire?

The quantum account gives a rational and naturalistic account of the correlation between mind and brain, by explaining it as a real understandable causal connection, based directly on the known laws of physics pertaining to the psychophysical connection, whereas the classical mechanistic account says that every physical connection can be explained without mentioning consciousness. But how, in this completely novel situation (of learning to use a prosthetic limb), does consciousness itself enter in a way that gives the illusion that it is playing a crucial causal role in the physical process when it is really doing nothing at all. Quantum theory, on the other hand, *needs* something to fill a specific causal gap, and provides the means for mind to fill it, whereas physically incorrect classical theory has no need for mind, and provides no means for it to do anything at all.

[Mondor] Dr. Stapp, thank you for your response. I see now that your use of the term 'free choice' has little or nothing to do with the traditional meaning(s) of 'free will' in philosophical literature. I am pleased to know that your theory does not require the existence of free will.

HPS: Yes, I certainly do not subscribe to the notion that our conscious choices have no causes or reasons whatever. But a commitment to the idea that each conscious choice has some reason to be what it turns out to be certainly does not entail that this reason can be specified completely in term of the localized physical variables of classical physics, or their direct quantum counterparts.

[Mondor] May I also assume it does not require consciousness, since it may be that the 'free choice' comes from unconscious mental activity preceding conscious awareness of it?

HPS: I repeatedly use the words 'orthodox', 'Copenhagen', or 'conventional' to emphasize the fact that I am describing the quantum theory that is used in actual scientific practice, and validated empirically.

As Wigner (1961b) (*Remarks on the Mind–Body Question*, see Wheeler and Zurek, p. 169) said: "It was not possible to formulate the laws of quantum mechanics in a completely consistent way without reference to the consciousness." (See Appendices C and D.) Also Heisenberg (1958a, p. 100): "The laws of nature that we formulate mathematically in quantum theory deal no longer with the particles themselves but with our knowledge of the elementary particles, [...] no longer the behavior of the elementary particles but rather our knowledge of this behavior."

Thus reference to the contents of our conscious streams of consciousness is an essential aspect of *conventional* quantum mechanics, and the basis of its testability and practical usability.

Some physicists (e.g., Bohm, and Ghirardi, Rimini, and Weber – and Philip Pearle) have tried to eliminate consciousness but all have failed to accommodate satisfactorily within their frameworks the fully relativistic QM with particle production (see Chap. 10). Others have tried the many-worlds approach, which many have noted ought really be called the 'one-world, many-minds' theory because it attempts to tie theory to the empirical/experiential data by trying to understand why the experienced world is so tremendously different from the 'one world' that would arise from using only the currently known dynamical laws that make no reference to consciousness – namely process 2 – by assuming (without specifying how) that this one objectively existing quantum world is experienced (subjectively) in myriads of different ways that hang together in countless separate streams of consciousness that individually manifest the statistical regularities specified by the quantum laws. To even begin to face the problems one must bring in the concept of consciousness. Then there is the question of whether and how the laws that generate these fantastic regularities in our streams of consciousness can be generated without bringing consciousness into the dynamics. Orthodox quantum mechanics achieves the matching to data only by means of an intricate theory of observations, which brings in vector spaces and basis vectors tied to possible conscious experiences, and a closely linked process 1.

The huge disparity between the structure of human experience and the structure generated by the purely physically described process 2 makes the discussion of the relationship between conscious experiences and physically described laws the primary issue in the use and understanding of quantum theory.

[Polanik] Enduring insight: the distinction between a property and that which has or exhibits that property.

HPS: Conscious experiences belong to streams of conscious experiences, and these streams are part of nature's process. This process has aspects that are described in psychological language: in terms of thoughts, ideas, or feelings that have various experiential qualities that have been given names by persons. Each person constitutes an aspect of nature's process that features a stream of consciousness. One can properly say, therefore, that a stream of conscious thoughts has psychologically described properties. What 'has' a psychologically described property is primarily a conscious experience; secondarily the stream of conscious experiences to which the experience belongs; and thirdly, the person (an aspect of nature's process) of which this stream of consciousness is an important aspect.

[Polanik] The next question is: If there is experience occurring, then to what is that experience occurring?

HPS: To what? I guess the correct question is: In what? And the answer is: In a stream of consciousness!

[Polanik] [...] it follows that there is something real to which experiences occur.

HPS: The idea that there is some physical structure 'to which experiences occur' goes far beyond what science says. Nor does science tell us that there is some immaterial entity 'to which experiences occur'. Each experience occurs in a stream of conscious, which is an aspect of nature's psychophysical process. It appears 'to' a 'person' only by virtue of the fact that a person is, *actually*, according to this theory, a stream of psychophysical events, and each experience – of the kind we are considering – belongs to some such stream. The verbal statement that an experience 'occurs to' the stream to which it belongs, suggests an ontological separation that the theory does not entail or embrace.

[Mondor] But in 1952 David Bohm published an interpretation of QM that [...] was completely deterministic.

HPS: But, in spite of massive intense effort, this result (a completely deterministic model) has not been able to be carried over to our premier theory, quantum electrodynamics, where particle creation and annihilation becomes important. And Bohm himself, when trying to generalize his model to include consciousness, was led to an infinite tower of guiding fields each being guided by a higher one (Bohm 1990). Logical closure was thereby lost.

[Edwards] When I say von Neumann believed in fairies I do not do so lightly. Fairies are supernatural beings the existence of which we can neither observe nor infer. Abstract egos seem to be that.

HPS: The account I have given above about the place of consciousness in nature, and in physics, is essentially my understanding of von Neumann. I find no fairies there. The term 'abstract ego' highlights the fact the process 1 choices must be fixed by causes that go beyond what the physically described process 2 can do. Insofar as the choice of the process 1 event is causally understandable in terms of the existing features of the theory, this choice should be determined in terms of the physically described and psychologically described aspects of the postulated streams of psychophysical events, together with the physically described potentialities. The theory makes to reference to any disembodied streams of consciousness, and entails no evident need for any such systems to exist or to influence the flow of the embodied streams of psychophysical events that the contemporary physical theory recognizes and deals with.

[Edwards] So Henry Stapp's comments are just about observing. It is the choosing process which seems to me supernatural because it seems to obey laws that have nothing to do with known physics and as far as I can see is unverifiable. [...] not conscious realities, but non-physical chooser: that for me is the ghost!

HPS: I follow William James's dictum: "The thought itself is the thinker." I introduce no ghosts. No new kind of entity need be doing the choosing. The process that determines the choice could depend irreducibly only upon the psychologically and physically described aspects of the existing contemporary theory. We do not currently know the nature of the 'dark energy' that is pushing all parts of the universe apart. But that does not mean that fairies are doing the job.

[Mondor] Through a fortunate and fortuitous connection I am able to forward Dr. Basil Hiley's reply to some of Dr. Stapp's remarks.

[Basil Hiley] Sent: 07 November 2006 14:00. Subject: Reply to Henry Stapp's Comments on QM and Consciousness.

I find it very difficult to enter into a discussion that has been going on between two other parties. One misses the main thrust of the argument and often raises different points that may not be central to the discussion. However, I will comment on the paragraph:

> As (1961b) (*Remarks on the Mind–Body Question*, see Wheeler and Zurek, p. 169) said: "It was not possible to formulate the laws of Quantum Mechanics in a completely consistent way without reference to the consciousness." Also Heisenberg: "The laws of nature that we formulate mathematically in quantum theory deal no longer with the particles themselves but with our knowledge of the elementary particles, [...] no longer the behavior of the elementary particles but rather our knowledge of this behavior."

These quotes are certainly correct and even though Wigner and Heisenberg were outstanding physicists (incidentally I did have the privilege of discussing some of these issues with both these men) these are merely opinions.

HPS: As descriptions of the Copenhagen interpretation of quantum theory that is used in actual scientific practice these are more than just opinions. The mathematical formalism of quantum theory is construed, within that interpretation, as merely a procedure for making predictions about relationships between perceptions, and that is the justification given by Wigner for his assertion. (See Wigner in the cited reference.)

I, of course, am following (as did Wigner) von Neumann's extension in which the conscious events become the psychologically described aspects of psychophysical events whose physically described aspects are brain events. The events are now to be regarded as ontologically real mind-brain events, with von Neumann's dynamical rules still in place to ensure the retention of the crucial predictive power of the theory.

On the other hand, I acknowledged and stressed that there are opposing viewpoints, including prominently, the one of David Bohm.

[Hiley] They are opinions that have always troubled me. I find it difficult to reconcile them with the historic origins of quantum mechanics. Remember it all started from our inability to explain the distribution of blackbody radiation and the stability of matter in terms of classical physics. Without the stability of matter there would be no life forms in which consciousness could be exhibited (not even Hoyle's Black Cloud). To use consciousness to formulate the laws of quantum mechanics seems circular, unless of course you assume some kind of universal consciousness lying at the centre of being as is proposed by certain forms of Hinduism.

HPS: Three different issues must be distinguished here. One is the question of how human consciousness enters into the dynamics of human brains *as these brains exist today*. This is the immediate subject of von Neumann's application of empirically validated quantum theory to mind–brain dynamics. And it is a problem in contemporary neuroscience. The second issue is: what caused the laws of nature to be what they are, with their incredible suitability for life – including stability and, more impressively, the fact that they support the fantastic properties of DNA. This is a very interesting question? Are there, as string theory is now supposed to entail, some 10^{500} possible worlds which could all exist, so that some of them could accidentally have all of the properties that exist in our universe? If so, then that is why stable matter, and life as we know it, exists in our universe: the highly improbable becomes highly probable just because the number of instances to consider is effectively infinite. The third issue is how did the nature and the involvement of consciousness evolve or change over the course of the development in our universe of our human brains. More generally, why does consciousness exist in association with certain kinds of physical systems – or exist at all.

Hiley claims that "to use consciousness to formulate the laws of quantum mechanics seems circular". His point, I think, is that consciousness has physical prerequisites, which must come into being before consciousness. Hence the laws that cause, or allow, the physical prerequisites to come into being should not depend on a consciousness that comes into being only later.

On the other hand, the laws must provide the potentiality for experiences to occur, when the conditions are right. We know that experiences of the kind that we know do in fact occur, and are tied in a mathematically beautiful and highly tested way to a partitioning process that allows the continuous smear generated by process 2 to produce the discrete 'closed indivisible phenomena' exhibiting 'the element of wholeness, symbolized by the quantum of action, and completely foreign to classical physical principles'. Given the extreme mathematical elegance of this aspect of quantum theory, it is unreasonable to think that it is somehow accidental, and dependent upon the arrival of human beings. The 'theory of observation' that we use so successfully must be an aspect of nature's process that far transcends its relevance to human consciousness.

[Hiley] Most physicists would expect to account for the stability of matter in a way that is independent of consciousness and certainly of human consciousness.

HPS: Stability is associated with process 2; consciousness with processes 1 and 3.

[Hiley] HPS: "Some physicists (e.g., Bohm, and Ghirardi, Rimini, and Weber – and Philip Pearle) have tried to eliminate consciousness but all have failed to accommodate relativistic QM with particle production" is just not correct. You can do it. The mathematics gets very messy but you can do it. The Dirac field has proved difficult but some of the results of the work by Lasenby and some of my more recent work show that this is now possible but much is left to be done.

HPS: The admission that 'much is left to be done' is worrisome. My reasons for believing this has not yet been achieved are given in Chap. 10.

The fact that there is even a potentiality for consciousness to arise – when the physical conditions are right – means that the nature of reality cannot be fundamentally the sort of reality conceived in classical physics, consisting wholly of totally mindless objects and fields, with no seed of, or hint of, or toe-hold for, consciousness. And quantum mechanics informs us that that even the purely physically described aspects of nature are not adequately conceptualized in terms of the qualities assigned to rocks by classical physics. In quantum theory the purely physically described aspects are mere *potentialities* for real events to occur. A potentiality is more like an idea than a persisting material substance, and it is treated in the theory as 'an idea of what might happen'.

Objective reality thus appears to be is suffused with idea-like qualities, both at the level of physically described 'objective potentialities' and at the level of the psychophysical happenings. These idea-like qualities eventually get tied to human conscious experiences, but they seem to be built into the basic structure of quantum theory itself, as a theory of idea-like potentialities for changes in idea-like potentialities.

In the context of our human experience, the net effect of the complex process described by the quantum dynamics seems to be to recreate events similar to ones that had in the past, under similar circumstances, created structures that preserved and extended order.

The direction of the advance from classical physics to quantum physics suggests that idea-like aspects of nature are not incidental or accidental, but are important features of a natural process that has a tendency to preserve and extend recognized order.

16 Impact of Quantum Mechanics on Human Values

Philosophers have tried doggedly for three centuries to understand the role of mind in the workings of a brain conceived to function according to principles of classical physics. We now know no such brain exists: no brain, body, or anything else in the real world is composed of those tiny bits of matter that Newton imagined the universe to be made of. Hence it is hardly surprising that those philosophical endeavors were beset by enormous difficulties, which led to such positions as that of the 'eliminative materialists', who hold that our conscious thoughts must be eliminated from our scientific understanding of nature; or of the 'epiphenomenalists', who admit that human experiences do exist, but claim that they play no role in how we behave; or of the 'identity theorists', who claim that each conscious feeling is exactly the same thing as a motion of particles that nineteenth century science thought our brains, and everything else in the universe, were made of, but that twentieth century science has found not to exist, at least as they were formerly conceived. The tremendous difficulty in reconciling consciousness, as we know it, with the older physics is dramatized by the fact that for many years the mere mention of 'consciousness' was considered evidence of backwardness and bad taste in most of academia, including, incredibly, even psychology and the philosophy of mind.

What you are, and will become, depends largely upon your values. Values arise from self-image: from what you believe yourself to be. Generally one is led by training, teaching, propaganda, or other forms of indoctrination, to expand one's conception of the self: one is encouraged to perceive oneself as an integral part of some social unit such as family, ethnic or religious group, or nation, and to enlarge one's self-interest to include the interests of this unit. If this training is successful your enlarged conception of yourself as good parent, or good son or daughter, or good Christian, Muslim, Jew, or whatever, will cause you to give weight to the welfare of the unit as you would your own. In fact, if well conditioned you may give more weight to the interests of the group than to the well-being of your bodily self.

In the present context it is not relevant whether this human tendency to enlarge one's self-image is a consequence of natural malleability, instinctual tendency, spiritual insight, or something else. What is important is that we human beings do in fact have the capacity to expand our image of 'self', and that this enlarged concept can become the basis of a drive so powerful that it becomes the dominant determinant of human conduct, overwhelming every other factor, including even the instinct for bodily survival.

But where reason is honored, belief must be reconciled with empirical evidence. If you seek evidence for your beliefs about what you are, and how you fit into Nature, then science claims jurisdiction, or at least relevance. Physics presents itself as the basic science, and it is to physics that you are told to turn. Thus a radical shift in the physics-based conception of man from that of an isolated mechanical automaton to that of an integral participant in a non-local holistic process that gives form and meaning to the evolving universe is a seismic event of potentially momentous proportions.

The quantum concept of man, being based on objective science equally available to all, rather than arising from special personal circumstances, has the potential to undergird a universal system of basic values suitable to all people, without regard to the accidents of their origins. With the diffusion of this quantum understanding of human beings, science may fulfill itself by adding to the material benefits it has already provided a philosophical insight of perhaps even greater ultimate value.

This issue of the connection of science to values can be put into perspective by seeing it in the context of a thumb-nail sketch of history that stresses the role of science. For this purpose let human intellectual history be divided into five periods: traditional, modern, transitional, post-modern, and contemporary.

During the 'traditional' era our understanding of ourselves and our relationship to Nature was based on 'ancient traditions' handed down from generation to generation: 'Traditions' were the chief source of wisdom about our connection to Nature. The 'modern' era began in the seventeenth century with the rise of what is still called 'modern science'. That approach was based on the ideas of Bacon, Descartes, Galileo and Newton, and it provided a new source of knowledge that came to be regarded by many thinkers as more reliable than tradition.

The basic idea of 'modern' science was 'materialism': the idea that the physical world is composed basically of tiny bits of matter whose contact interactions with adjacent bits completely control everything that is now happening, and that ever will happen. According to these laws, as they existed in the late nineteenth century, a person's conscious thoughts and efforts can make no difference at all to what his body/brain does: whatever you do was deemed to be completely fixed by local interactions between tiny mechanical elements, with your thoughts, ideas, feelings, and efforts, being simply locally determined high-level consequences or re-expressions of the low-level mechanical process, and hence basically just elements of a reorganized way of describing the effects of the absolutely and totally controlling microscopic material causes.

This materialist conception of reality began to crumble at the beginning of the twentieth century with Max Planck's discovery of the quantum of action. Planck announced to his son that he had, on that day, made a discovery as important as Newton's. That assessment was certainly correct: the ramifications of Planck's discovery were eventually to cause Newton's materialist conception of physical reality to come crashing down. Planck's discovery marks the beginning of the 'transitional' period.

A second important transitional development soon followed. In 1905 Einstein announced his special theory of relativity. This theory denied the validity of our intuitive idea of the instant of time 'now', and promulgated the thesis that even the most basic quantities of physics, such as the length of a steel rod, and the temporal order of two events, had no objective 'true values', but were well defined only 'relative' to some observer's point of view.

Planck's discovery led by the mid-1920s to a complete breakdown, at the fundamental level, of the classical material conception of nature. A new basic physical theory, developed principally by Werner Heisenberg, Niels Bohr, Wolfgang Pauli, and Max Born, brought 'the observer' explicitly into physics. The earlier idea that the physical world is composed of tiny particles (and electromagnetic and gravitational fields) was abandoned in favor of a theory of natural phenomena in which the consciousness of the human observer is ascribed an essential role. This successor to classical physical theory is called Copenhagen quantum theory.

This turning away by science itself from the tenets of the objective materialist philosophy gave impetus to, and lent support to, postmodernism. That view, which emerged during the second half of the twentieth century, promulgated, in essence, the idea that all 'truths' were relative to one's point of view, and were mere artifacts of some particular social group's struggle for power over competing groups.

Thus each social movement was entitled to its own 'truth', which was viewed simply as a socially created pawn in the power game.

The connection of post-modern thought to science is that both Copenhagen quantum theory and relativity theory had retreated from the idea of observer-independent objective truth. Science in the first quarter of the twentieth century had not only eliminated materialism as a possible foundation for objective truth, but seemed to have discredited the very idea of objective truth in science. But if the community of scientists has renounced the idea of objective truth in favor of the pragmatic idea that 'what is true for us is what works for us', then every group becomes licensed to do the same, and the hope evaporates that science might provide objective criteria for resolving contentious social issues.

This philosophical shift has had profound social and intellectual ramifications. But the physicists who initiated this mischief were generally too interested in practical developments in their own field to get involved in these philosophical issues. Thus they failed to broadcast an important fact: already by mid-century, a further development in physics had occurred that provides an effective antidote to both the 'materialism' of the modern era, and the 'relativism' and 'social constructionism' of the post-modern period. In particular, John von Neumann developed, during the early thirties, a form of quantum theory that brought the physical and mental aspects of nature back together as two aspects of a rationally coherent whole. This theory was elevated, during the forties – by the work of Tomonaga and Schwinger – to a form compatible with the physical requirements of the theory of relativity.

Von Neumann's theory, unlike the transitional ones, provides a framework for integrating into one coherent idea of reality the empirical data residing in subjective experience with the basic mathematical structure of theoretical physics. Von Neumann's formulation of quantum theory is the starting point of all efforts by physicists to go beyond the pragmatically satisfactory but ontologically incomplete Copenhagen form of quantum theory.

Von Neumann capitalized upon the key Copenhagen move of bringing human choices into the theory of physical reality. But, whereas the Copenhagen approach excluded the bodies and brains of the human observers from the physical world that they sought to describe, von Neumann demanded logical cohesion and mathematical precision, and was willing to follow where this rational approach led. Being a mathematician, fortified by the rigor and precision of his thought, he seemed less intimidated than his physicist brethren by the sharp contrast between the nature of the world called for by the new mathematics and the nature of the world that the genius of Isaac Newton had concocted.

A common core feature of the orthodox (Copenhagen and von Neumann) quantum theory is the incorporation of efficacious conscious human choices into the structure of basic physical theory. How this is done, and how the conception of the human person is thereby radically altered, has been spelled out in lay terms in this book, and is something every well informed person who values the findings of science ought to know about. The conception of self is the basis of values and thence of behavior, and it controls the entire fabric of one's life. It is irrational, from a scientific perspective, to cling today to false and inadequate nineteenth century concepts about your basic nature, while ignoring the profound impact upon these concepts of the twentieth century revolution in science.

It is curious that some physicists want to improve upon orthodox quantum theory by excluding 'the observer', who, by virtue of his subjective nature, must, in their opinion, be excluded from science. That stance is maintained in direct opposition to what would seem to be the most profound advance in physics in three hundred years, namely the overcoming of the most glaring failure of classical physics, its inability to accommodate us, its creators. The most salient philosophical feature of quantum theory is that the mathematics has a causal gap that, by virtue of its intrinsic form, provides a perfect place for Homo sapiens as we know and experience ourselves.

17 Conclusions

How can our world of billions of thinkers ever come into general concordance on fundamental issues? How do you, yourself, form opinions on such issues? Do you simply accept the message of some 'authority', such as a church, a state, or a social or political group? All of these entities promote concepts about how you as an individual fit into the reality that supports your being. And each has an agenda of its own, and hence its own internal biases. But where can you find an unvarnished truth about your nature, and your place in Nature?

Science rests, in the end, on an authority that lies beyond the pettiness of human ambition. It rests, finally, on stubborn facts. The founders of quantum theory certainly had no desire to bring down the grand structure of classical physics of which they were the inheritors, beneficiaries, and torch bearers. It was stubborn facts that forced their hand, and made them reluctantly abandon the two-hundred-year-old classical ideal of a mechanical universe, and turn to what perhaps should have been seen from the start as a more reasonable endeavor: the creation an understanding of nature that includes in a rationally coherent way the thoughts by which we know and influence the world around us. The labors of scientists endeavoring merely to understand our inanimate environment produced, from its own internal logic, a rationally coherent framework into which we ourselves fit neatly. What was falsified by twentieth-century science was not the core traditions and intuitions that have sustained societies and civilizations since the dawn of mankind, but rather an historical aberration, an impoverished world view within which philosophers of the past few centuries have tried relentlessly but fruitlessly to find ourselves. The falseness of that deviation of science must be made known, and heralded, because human beings are not likely to endure in a society ruled by a conception of themselves that denies the essence of their being.

A Gazzaniga's *The Ethical Brain*

Michael S. Gazzaniga is a renowned cognitive neuroscientist. He was Editor-in-Chief of the 1447 page book *The Cognitive Neurosciences*, which, for the past decade, has been the fattest book in my library, apart from 'the unabridged'. His recent book *The Ethical Brain* has a Part III entitled *Free Will, Personal Responsibility, and the Law*. This part addresses, from the perspective of cognitive neuroscience, some of the moral issues that have been dealt with in the present book. The aim of his Part III is to reconcile the materialist idea that brain activity is *determined* with the notion of *moral responsibility*, which normally depends upon the idea that we human beings possess free will.

Gazzaniga asserts:

> Based on the modern understanding of neuroscience and on the assumptions of legal concepts, I believe the following axioms: Brains are automatic, rule-governed, determined devices, while people are personally responsible agents, free to make their own decisions.

One possible interpretation of these words – the quantum-theoretic interpretation – would be that a person has both a mind (his stream of conscious thoughts, ideas, and feelings) and a brain (made of neurons, glia, etc), and that his decisions (his conscious moral choices) are free (not determined by any known law), and that, moreover, the rules that govern his brain *determine* the activity of his brain *jointly* from the physically described properties of the brain *combined* with these conscious decisions. That interpretation is essentially what orthodox (von Neumann) quantum mechanics – and also common sense intuition – asserts.

If this interpretation is what Gazzaniga means, then there is no problem. But I believe that this is *not* what Gazzaniga means. Earlier on he said:

> The brain determines the mind, and the brain is a physical entity subject to all the rules of the physical world. The physical world is determined, so our brains must also be determined.

This seems to be suggesting that by 'determined' he means determined solely by physically described properties, as would be the case if the concepts of classical physics were applicable. However, what he actually said was that "the brain is a physical entity subject to all the rules of the physical world". The rules of the physical world, as specified by contemporary (orthodox quantum) theory, explain how the brain is governed in part by the brain and in part by our conscious choices, which themselves are not governed by any known laws. If this physics-based understanding of 'determined' is what Gazzaniga means then there is no difficulty in reconciling the fact that an agent's brain is 'determined' with the fact that this agent's person is 'free': the agent's *brain* is determined partly by his brain and partly by his conscious free choices, and hence the *person* whose actions this brain controls is likewise jointly controlled by these two factors, neither of which alone suffices.

If this contemporary-physics-based interpretation is what Gazzaniga meant, then he could have stopped his book right there: that interpretation is in complete accord with common sense, with normal ethical theory, and with contemporary physics. Thus the fact that he did not stop, but went on to write his book, including Part III, suggests that he is using not the quantum mechanical meaning of 'determined'; but rather the meaning that would hold in the classical approximation, which exorcizes all the physical effects of our conscious choices. Indeed, he goes on to say:

> If our brains are determined, then [...] is the free will we seem to experience just an illusion? And if free will is an illusion, must we revise our concepts of what it means to be personally responsible for our actions?

I am assuming in this appendix that Gazzaniga is adhering essentially to nineteenth century physics, so that 'determined' means automatically/mechanically determined by physically described properties alone, like a clock, and that he is thus endeavoring to address the question: How can one consider a person with an essentially clock-like body-brain to be morally responsible for his actions? How can we uphold the concept of ethical behavior within the confines of an understanding of nature that reduces each human being to a mechanical automaton?

Gazzaniga's answer is built upon a proposed restructuring (redefining) the meanings of both 'free will' and 'moral responsibility'. Following an idea of David Hume, and more recently of A.J. Ayer, the word 'free' is effectively defined to mean 'unconstrained by external bonds'. Thus a clock is 'free' if the movements of its hands and cogs are not restricted by external bonds or forces. However, the 'free will' of traditional ethical theory refers to a type of freedom that a mechanically controlled clock would *not* enjoy, even if it had no *external* bonds. This latter – morally pertinent – kind of free will is specifically associated with consciousness. Thus a physically determined clock *that has no consciousness is not subject* to moral evaluation, even if it is not constrained by external bonds, whereas a person possessing a conscious 'will' that is physically efficacious, yet not physically determined, *is subject* to moral evaluation when he is not constrained by external bonds. Thus the morally pertinent idea of 'possessing free will' is not the same as 'unconstrained by external bonds or forces'. The Hume/Ayer move obscures the morally pertinent idea of freedom, which is intimately linked to consciousness, by confounding it with different idea that does not specifically involve consciousness. This move throws rational analysis off track by suppressing (on the basis of an inapplicable approximation) the involvement of consciousness in the morally relevant conception of 'free will'.

Ethical and moral values traditionally reside in the ability of a person to make discerning conscious judgments pertaining to moral issues, coupled with the capacity of the person's conscious effort to willfully force his body to act in accordance with the standards he has consciously judged to be higher, in the face of strong natural tendencies to do otherwise. The whole moral battle is fought in the realm of conscious thoughts, ideas, and feelings. Where there is no consciousness there is no moral dimension. *Moreover*, if consciousness exists but is permitted by general rules to make no physical difference – that is, if consciousness is constrained by the general laws to be an impotent witness to mechanically determined process – then the seeming struggle of will becomes a meaningless charade, and the moral dimension again disappears.

It is the imposition, by virtue of the classical approximation, of this law-based kind of impotency that eliminates the moral dimension within that approximation. The morally pertinent free will is eradicated by the classical approximation even if there are no external bounds. Calling a system 'free' just because it is *not constrained by external bonds* does not suffice to give that system the kind of free will that undergirds normal ethical ideas.

Gazzaniga's attack on the problem has also a second prong. He avers that: "Personal responsibility is a public concept." He says of things such as personal responsibility that:

> Those aspects of our personhood are – oddly – not in our brains. They exist *only* in the relationships that exist when our automatic brains interact with other automatic brains. They are in the ether.

This idea that these pertinent things are "in the ether" and exist "only in the relationships" is indeed an *odd* thing for a materialistically-oriented neuroscientist to say. It seems mystical. Although ideas about personal responsibility may indeed arise only in social contexts, one would normally say that the resulting ideas about personal responsibility *exist in the streams of consciousness of the interacting persons*, and a materialist would be expected to say that these ideas are 'in' or are 'some part of' the brains of those socially interacting persons. Yet if the causes of self-controlled behavior are wholly in the brains and bodies of the agents, and these brains and bodies are automatically determined by the physically described body-brain alone, then it is hard to see how these agents, as persons, can have the kind of free will upon which our moral and ethical theories are based. Some sort of odd or weird move is needed to endow a person with morally relevant free will if his body and brain are mechanically determined.

But if some sort of weirdness is needed to rescue the social concept of personal responsibility, then why not use 'quantum weirdness'. The quantum concepts may seem weird to the uninitiated, but they are based on science, and they resolve the problem of moral responsibility by endowing our conscious choices with causal influence in the selection of our physical actions.

It is hard to see the advantage of introducing the changes described by Gazzaniga compared to the option of simply going beyond the in-principle-inadequate classical approximation. Why do thinkers dedicated to rationality resist so tenaciously the option of accepting (contemporary orthodox quantum) physics, which says that our conscious choices intervene, in a very special and restricted kind of way, in the mechanically determined time development of the physically described aspects of a system – during the process by means of which the conscious agent acquires new knowledge about that system? Because acquiring new knowledge about a system normally involves a probing of the system, it is not at all weird that the system being examined should be affected by the extraction of knowledge from it, and hence comes to depend upon how it was probed.

The advantages of accepting quantum mechanics in cognitive neuroscience, and ultimately in our lives, are:

- It is compatible with basic physical theory, and thus will continue to work in increasingly complex and miniaturized empirical situations.
- It specifies how a person's consciously experienced intentional choices are represented in the physically described aspects of the theory.
- It removes the incoherency of a known-to-be-real ontological element that contains the empirical data, yet resides in a realm that has no law-based connection to the flow of physical events.
- It provides a foundation for understanding the co-evolution of mind and brain, because each of these two parts contributes to the dynamics in a way that is linked to the other by laws that are specified, at least in part.
- It provides for a free will of the kind needed to undergird ethical theory.
- It produces a science-based image of oneself, not as a freak-accident out-cropping with consciousness riding like a piece of froth on the ocean – but rather as an active component of a deeply interconnected world process that is responsive to value-based human judgments.

B Von Neumann: Knowledge, Information, and Entropy

The book *John von Neumann and the Foundations of Quantum Physics* (Rédei 2001) contains a fascinating and informative article written by Eckehart Köhler entitled *Why von Neumann Rejected Carnap's Dualism of Information Concept*. The topic is precisely the core issue before us: How is knowledge connected to physics? Köhler illuminates von Neumann's views on this subject by contrasting them to those of Carnap.

Rudolph Carnap was a distinguished philosopher, and member of the Vienna Circle. He was in some sense a dualist. He had studied one of the central problems of philosophy, namely the distinction between *analytic* statements and *synthetic* statements. (The former are true or false by virtue of a specified set of rules held in our minds, whereas the latter are true or false by virtue their concordance with physical or empirical facts.) His conclusions had led him to the idea that there are two different domains of truth, one pertaining to logic and mathematics and the other to physics and the natural sciences. This led to the claim that there are 'two concepts of probability', one logical the other physical. That conclusion was in line with the fact that philosophers were then divided between two main schools as to whether probability should be understood in terms of abstract idealizations or physical sequences of outcomes of measurements. Carnap's bifurcations implied a similar division between two different concepts of information, and of entropy.

In 1952 Carnap was working at the Institute for Advanced Study in Princeton and about to publish a work on his dualistic theory of information, according to which epistemological concepts like information should be treated separately from physics. Von Neumann, in private discussion, raised objections, and Pauli later wrote a forceful letter, asserting that: "I am quite strongly opposed to the position you take." Later he adds: "I am indeed concerned that the confusion in the area of the foundations of statistical mechanics not grow further (and I fear very much that a publication of your work in its present form would have this effect)."

Carnap's view was in line with the Cartesian separation between a domain of real objective physical facts and a domain of ideas and concepts. But von Neumann's view, and also Pauli's, linked the probability that occurred in physics, in connection with entropy, to *knowledge*, in direct opposition to Carnap's view that epistemology (considerations pertaining to knowledge) should be separated from physics. The opposition of von Neumann and Pauli significantly delayed the publication of Carnap's book.

This issue of the relationship of knowledge to physics is the central question before us, and is in fact the core problem of all philosophy and science. In the earlier chapters I relied upon the basic insight of the founders of quantum theory, and upon the character of quantum theory as it is used in actual practice, to justify the key postulate that Process 1 is associated with knowing, or feeling. But there is also an entirely different line of justification of that connection developed in von Neumann's book, *Mathematical Foundations of Quantum Mechanics*. This consideration, which strongly influenced his thinking for the remainder of his life, pertains to the second law of thermodynamics, which is the assertion that entropy (disorder, defined in a precise way) never decreases.

There are huge differences in the quantum and classical workings of the second law. Von Neumann's book discusses in detail the quantum case, and some of those differences. In one sense there is no nontrivial objective second law in classical physics: a classical state is supposed to be objectively well defined, and hence it always has probability one. Consequently, the entropy is zero at the outset and remains so forevermore. Normally, however, one adopts some rule of 'coarse graining' that destroys information and hence allows probabilities to be different from unity, and then embarks upon an endeavor to deduce the laws of thermodynamics from statistical considerations. Of course, it can be objected that the subjective act of choosing some particular coarse graining renders the treatment not completely objective, but that limited subjective input seems insufficient to warrant the claim that physical probability is closely tied to knowledge.

The question of the connection of entropy to the *knowledge and actions of an intelligent being* was, however, raised in a more incisive form by Maxwell, who imagined a tiny 'demon' to be stationed at a small doorway between two large rooms filled with gas. If this agent could distinguish different species of gas molecules, or their energies and locations, and slide a frictionless door open or closed according to which type of molecule was about to pass, he could easily cause a decrease in entropy that could be used to do work, and hence to power a perpetual motion machine, in violation of the second law.

This paradox was examined Leo Szilard, who replaced Maxwell's intelligent demon by a simple idealized (classical) physical mechanism that consumed no energy beyond the apparent minimum needed to 'recognize and respond differently to' a two-valued property of the gas molecule. He found that this rudimentary process of merely 'coming to know and respond to' the two-valued property transferred entropy from heat baths to the gaseous system in just the amount needed to preserve the second law. Evidently nature is arranged so that what we conceive to be the purely intellectual process of coming to know something, and acting on the basis of that knowledge, is closely linked to the probabilities that enter into the constraints upon physical processes associated with entropy.

Von Neumann describes a version of this idealized experiment. Suppose a single molecule is contained in a volume V. Suppose an agent comes to know whether the molecule lies to the left or to the right of the center line. He is then in the state of being able to order the placement of a partition/piston at that line and to switch a lever either to the right or to the left, which restricts the direction in which the piston can move. This causes the molecule to drive the piston slowly to the right or to the left, and transfer some of its thermal energy to it. If the system is in a heat bath then this process extracts from the heat bath an amount 'log 2' of entropy (in natural units). Thus the *knowledge* of which half of the volume the molecule was in is converted into a decrement of 'log 2' units of entropy. In von Neumann's words: "We have exchanged our knowledge for the entropy decrease $k \log 2$." (Here k is the natural unit of entropy.)

What this means is this: When we conceive of an increase in the 'knowledge possessed by some agent' we must not imagine that this knowledge exists in some ethereal kingdom, apart from its physical representation in the body of the agent. Von Neumann's analysis shows that the change in knowledge represented by Process 1 is quantitatively tied to the probabilities associated with entropy.

Among the many things shown by von Neumann are these two:

1. The entropy of a system is unaltered when the state of that system is evolving solely under the governance of process 2.
2. The entropy of a system is never decreased by any process 1 event.

The first result is analogous to the classical result that if an objective 'probability' were to be assigned to each of a countable set of possible classical states, and the system were allowed to evolve in accordance with the classical laws of motion then the entropy of that system would remain fixed.

The second result is a nontrivial quantum second law of thermodynamics. Instead of coarse-graining, one has process 1, which in the simple 'Yes–No' case converts the prior system into one where the question associated with the projection operator P has a definite answer, but only the *probability* associated with each possible answer is specified, not an answer itself.

One sees, therefore, why von Neumann rejected Carnap's attempt to divorce knowledge from physics: large tracts in his book were devoted to establishing their marriage. That work demonstrates the quantitative link between the increment of knowledge or information associated with a process 1 event and the probabilities connected to entropy. This focus on process 1 allowed him to formulate and prove a quantum version of the second law. In the quantum universe the rate of increase of entropy would be determined not by some imaginary and arbitrary coarse graining rule, but by the number and nature of objectively real process 1 events.

Köhler discusses another outstanding problem: the nature of mathematics. At one time mathematics was imagined to be an abstract resident of some immaterial Platonic realm, independent in principle from the brains and activities of those who do it. But many mathematicians and philosophers now believe that the process of doing mathematics rests in the end on mathematical intuitions, which are essentially aesthetic evaluations.

Köhler argues that von Neumann held this view. But what is the origin or source of such aesthetic judgments?

Roger Penrose based his theory of consciousness on the idea that mathematical insight comes from a Platonic realm. But according to the present account each such illumination, like any other experience, is represented in the quantum description of nature as a picking out of an organized state in which diverse brain processes act together in an harmonious state of mutual support that leads on to feedbacks that sustain the structure by recreating it with slight variations, in the quantum state of the agent's brain/body. Every experience of any kind is fundamentally like this: it is a process 1 grasping of a state of order that tends to recreate itself in a slightly varied form.

This notion that each process 1 event is a felt grasping of a state in which various sub-processes act in concert to produce an ongoing continuation of itself provides a foundation for answering in a uniform way many outstanding philosophical and scientific problems. For example, it provides a foundation for a solution to a basic issue of neuroscience, the so-called 'binding problem'. It is known that diverse features of a visual scene, such as color, location, size, shape, etc., are processed by separate modules located in different regions of the brain. This understanding of the process 1 event makes the felt experience a grasping of a non-discordant quasi-stable mutually supportive combination of these diverse elements as a unified whole. To achieve maximal organizational impact this event should provide the conditions for a rapid sequence of re-enactments of itself. Then this conception of the operation of von Neumann's process 1 provides also an understanding of the capacity of an agent's thoughts to control its bodily behavior. The same conception of process 1 provides also a basis for understanding both artistic and mathematical creativity, and the evolution of consciousness in step with the biological evolution of our species. These issues all come down to the problem of the connection of knowings to physics, which von Neumann's treatment of entropy ties to process 1.

Köhler quotes an interesting statement of von Neumann, but then draws from it conclusions about von Neumann's views that go far beyond what von Neumann actually said.

Von Neumann points out that in classical mechanics one can solve the problem of motion either by solving differential equations (the local causal mechanistic approach) or by using a global least action (or some other similar) approach. This latter method can be viewed as 'teleological' in the sense that if initial and final conditions are specified then the principle of least action specifies the path between them. He goes on to say that he is:

> [...] not trying to be facetious about the importance of keeping teleological principles in mind when dealing with biology; but I think one hasn't started to understand the problem of their role in biology until one realizes that in mechanics, if you are just a little bit clever mathematically, your problem disappears and becomes meaningless. And it is perfectly possible that if one understood another area then the same thing might happen.

The pertinent 'other area' is psychology, or the problem of mind. The first point is that von Neumann's statement is very cautious: he says that it is "perfectly *possible* that *if* one understood another area the same thing *might* happen." There are three weak links: 'possible', 'if', and 'might'.

Köhler's conclusion is far less cautious. He follows the above quotation with the assertion:

> Here von Neumann warns biologists against overstressing goal-directed activity since this can always be reformulated *causally*.

Von Neumann said no such thing. He merely points out that in classical mechanics certain global least action principles are equivalent to local causal mechanistic rules. That falls far short of claiming that *all* goal-directed activity can be expressed in least-action terms, or that in *non-classical* cases such a least-action formulation would necessarily be equivalent to a local causal mechanism. Von Neumann recognizes this as a possibility, not a necessity.

In quantum physics the process 2 part of the dynamics is derived from the quantization of the classical law. Hence it might be contended that *for this process 2 part of the dynamics* an equivalence holds between 'teleological' and 'causal' formulations. But the connection to mind involves process 1. It is far from obvious that the equivalence found in classical mechanics will carry over to process 1. In the first place, process 1 involves non-local operators P, and that alone would appear to block reduction to local causation. In the second place, Process 1 drops out of the dynamics when one goes to the classical limit, which is the limit in which all effects involving Planck's constant are neglected. Hence process 1 is, in this sense, non-classical or anti-classical. Hence there is no reason to believe that equivalences occurring in classical physics will carry over to process 1. Such a connection 'might possibly' hold, but it is surely not required to hold by anything we know today.

Köhler goes on to state that:

> Based on his general approach, one may say von Neumann was a psychophysical reductionist who thought human intelligence could in principle be presented and explained on a physical level – in particular, neurologically, in terms of nerve nets. Between the physiology of nerves and the physics of computer devices von Neumann recognized no difference in functional capacity.

That last statement seems tremendously at odds with the conclusions of von Neumann's final work, *The Computer and the Brain*, which emphasized the huge differences between brains and computers. But, that point aside, the fact that von Neumann did much work on classically describable computers does not imply that he was committed to the view that *human intelligence* could be understood in classical terms. Von Neumann may indeed have not excluded that possibility, but I doubt that any statement of his shows him to be committed to the position that human intelligence, and, more importantly, his process 1, can be explained in local mechanistic terms. The statement quoted above certainly fails to justify such a conclusion.

C Wigner's Friend and Consciousness in Quantum Theory

Eugene Wigner published in 1961 a widely reprinted article (Wigner 1961b) entitled *Remarks on the Mind–Body Problem* in which he stresses the basic role played by consciousness in quantum theory. But if consciousness is basic then the question immediately arises: Whose consciousness? To explore this issue Wigner considers a situation in which his 'friend', rather than he himself, is observing the effects of an atomic process, the radiation of a visible photon.

In order to formulate the problem Wigner first explains the entry of consciousness into physical theory:

> When the province of physical theory was extended to encompass microscopic phenomena, through the creation of quantum mechanics, the concept of consciousness came to the fore again: it was not possible to formulate the laws of quantum mechanics without reference to the consciousness.3 All that quantum mechanics purports to describe are probability connections between subsequent impressions (also called 'apperceptions') of consciousness, and even though the dividing line between the observer, whose consciousness is being affected, and the observed physical object can be shifted towards one or the other to a considerable degree,[4] it cannot be eliminated.

His reference 4 [von Neumann 1932, Chap. VI] is to von Neumann's work [orthodox interpretation] on the shifting of the boundary between those aspects of nature that are described in the mathematical language of quantum theory, and those that are described in the psychological language by means of which we describe our actual and possible conscious experiences. The job of quantum theory is to make predictions about connections between such experiences. His reference 3 [Heisenberg 1958] was to Heisenberg's famous pronouncement:

> The conception of objective reality [...] evaporated into the [...] mathematics that represents no longer the behavior of elementary particles but rather our knowledge of this behavior.

The concept of 'our knowledge' is reasonably clear insofar as "we are able to communicate to others what we have done and what we have learnt" [Bohr 1962, p. 3]. But in practice different people often know different things.

The thought experiment considered by Wigner involves, essentially, an atomic state that emits a visible photon into an optical system that directs the rays emitted from the atom in certain directions into the retina of the eye of Wigner's friend, and directs the rays emitted in other directions to some other place. The wave function of the atom plus the photon will be a superposition of components corresponding to different directions of the photon emission. If the interaction of the photon with the retina, and of the retina with the brain of the friend – who is presumed to be attending to what she is seeing – is now included in the physical description, then the state of his friend's brain generated by the purely physical laws of motion would include a part that corresponds to her observing the flash and another part corresponding to her not observing the flash. When Wigner asks his friend whether she saw the flash, then, upon his registering of her response, the wave function (quantum state) that represents his knowledge of her brain and body will suddenly jumps to one state or the other. Yet before he learned about her reaction his representation of her state was in a combination of the 'I observed a flash' and 'I observed no flash' alternatives.

Wigner is willing to admit that, if the purely physically described laws entail it, then an unobserved inanimate measuring device could exist in a state that represents a combination of two macroscopically different states. However, he notes that although solipsism may be a *logical* possibility, "everyone believes that the phenomena of sensation are widely shared by organisms that we consider to be living". And, accordingly, his friend will surely report that she did (or did not) experience the flash (as the case may be) *before* she reported that fact to him. Wigner concludes from these considerations that his friend was "not in a state of suspended animation" before *he* learned about her state: he concludes that her quantum state became one or the other of these two alternatives when *she* became conscious of the flash, not when *he* came to know what she reported.

Wigner asserts that:

> The preceding argument for the difference in the roles of inanimate tools of observation and observers with consciousness – hence for a violation of physical laws where consciousness plays

a role – is entirely cogent so long as one accepts the tenets of orthodox quantum theory and all their consequences.

Wigner proposes, then, that "the being with a consciousness must have a different role in quantum mechanics than the inanimate measuring device". He proposes, in essence, that the occurrence of a conscious experience is an objective reality that is correlated to a change in an objective wave function. 'Our knowledge' can then be interpreted to be the aggregate of the conscious knowledge of *all* systems that possess consciousness. This allows quantum theory to be regarded as an objective theory that describes the interaction between an objective physical aspect that is described in terms of the mathematical language of quantum theory, and an objective mental aspect that is described in terms of the concepts of thoughts, ideas, and feelings – i.e., in terms of the concepts of psychology. This move allows what had originally been a fundamentally anthropocentric, pragmatic, subjective theory to be elevated into a non-anthropocentric objective theory of an objective reality having physically described aspects and psychologically described aspects, related in the way specified by the orthodox interpretation quantum theory spelled out by John von Neumann (1932).

D Orthodox Interpretation and the Mind–Brain Connection

Eugene Wigner, in a paper entitled *The Problem of Measurement* (Wigner 1963), used the term 'orthodox interpretation' to identify the interpretation spelled out in mathematical detail by John von Neumann in his book *Mathematische Grundlagen der Quantnmechanik* (von Neumann 1932). Von Neumann, in the chapter on the measuring process, shows how to expand the quantum mechanical description of a system to include the physical variables of the measuring device, or, more generally, the physical variables of any system that interacts with an original system of interest. He then gives a detailed analysis of the process of measurement.

Von Neumann calls the unitary evolution of the quantum state (or wave function) generated by the Schroedinger equation by the name 'process 2'. The process 2 quantum mechanical evolution is a mathematical generalization of the deterministic evolution of a dynamically closed system in classical physical theory. The quantum mechanical process 2, like its classical counterpart, is deterministic: given the quantum state at any time, the state into which will evolve at any later time via process 2 is completely fixed.

Von Neumann considers an (idealized) situation involving a sequence of physically described measuring devices each performing a good measurement on the outcome variables of the preceding device, leading eventually to the retina, then to the optical nerves, and finally to the higher brain centers directly associated with the consciousness of the observer. There is no apparent reason for the process 2 to fail at any point, provided the full environment (essentially the entire physically described universe) is included in the physical system. But in general the process 2 evolution will lead to a state in which the higher brain centers directly associated with consciousness will have non-negligible components corresponding to different incompatible experiences, such as seeing the pointer of a measuring device simultaneously at several distinct positions.

Von Neumann (1955/1932, p. 418) notes that:

> It is entirely correct that the measurement or the related process of subjective perception is a new entity relative to the physical environment and is not reducible to the latter. Indeed, it leads into the intellectual inner life of the individual, which is extra-observational by its very nature (since it must be taken for granted by any conceivable observation or experiment).

To tie the quantum mathematics usefully to human experience von Neumann invokes another process, which he called 'process 1'. Process 1 partitions the state into a particular collection of components each corresponding to a distinct possible experience, but only one of which will survive the 'collapse of the wave function' or the 'reduction of the wave packet' associated with the process of observation.

Wigner proves that process 1 can never be a consequence of process 2 alone: some other process, not the quantum analog of the deterministic classical law of evolution, must come in. As in the classical case, one must of course respect the condition that the quantum system be dynamically closed. This means that if any macroscopic element is included in the quantum mechanically described system then one must effectively include the whole universe, due to the non-negligible effects of the interaction between a macroscopic system and its environment.

Von Neumann notes that, in line with the precepts of the Copenhagen interpretation:

> [...] we must always divide the world into two parts, the one being the observed system, the other the observer,

and that

> [...] quantum mechanics describes the events which occur in the observed portion of the world, so long as they do not interact with the observing portion, with the aid of process 2, but as soon as such an interaction occurs, i.e., a measurement, it requires an application of process 1.

The von Neumann/Wigner approach is, in this regard, not identical to the Copenhagen interpretation specified by Bohr and Heisenberg, who, in keeping with their pragmatic epistemological stance, resist treating the entire physical universe as a quantum system obeying the linear deterministic unitary law. Bohr ties this limitation in the applicability of the normal quantum rules to the fact that any attempt to obtain sufficient knowledge about any living organism, in order to enable us to make useful predictions, would probably kill the organism. Hence "the strict application of those concepts adapted to our description of inanimate nature might stand in a relationship of exclusion to the consideration of the laws of the phenomena of life" (Bohr 1961, pp. 22–23). This argument is effectively a cautious suggestion that the breakdown of process 2 might be associated with biological systems, i.e., with life. But von Neumann says:

> There arises the frequent necessity of localizing some of these processes at points which lie within the portion of space occupied by our own bodies. But this does not alter the fact of their belonging to the 'world about us', the objective environment referred to above.

Wigner's suggestion for dealing with this gross mismatch between the process-2-generated activities of our brains and the contents of our streams of conscious experiences, evidently stems from a desire to have a rationally coherent *ontological* understanding of nature herself; an understanding of the reality that actually exists. Noting that process 1 is associated with the occurrence of *observable* events, and hence the implied need for an observer, Wigner suggest that the breakdown of process 2 is due to the interaction of the physically described aspects of nature with the consciousness of a conscious being (Wigner 1961). This physically efficacious consciousness stands outside the physically described aspects of nature controlled by process 2. Von Neumann calls it the observer's 'abstract ego'.

Conscious experiences are certainly real, and real things normally have real effects. The most straightforward conclusion would seem to be that process 1 specifies features of the interaction between (1), the brain activities that are directly associated with conscious experiences, and (2), the conscious experiences with which those activities are associated.

This solution is in line with Descartes' idea of two 'substances', that can interact in our brains, provided 'substance' means merely a carrier of 'essences' The essence of the inhabitants of res cogitans is 'felt experience'. They are thoughts, ideas, and feelings: the realities that hang together to form our streams of conscious experiences. But the essence of the inhabitants of res extensa is not at all that of the sort of persisting stuff that classical physicists imagined the physical world to be made of. These properties are indeed represented in terms of mathematically described properties assigned to spacetime points, but their essential nature is that of "potentialities for the psychophysical events to occur". These events occur at the interface between the psychologically and physically described aspects of nature, and the laws governing their interaction are given by von Neumann. The causal connections between "potentialities for psychologically described events to occur" and the actual occurrence of such events are easier to comprehend and describe than causal connections between the mental and physical features of classical physics. For, both sides of the quantum duality are conceptually more like 'ideas' than like 'rocks'.

E Locality in Physics

In physics there is a condition, or at least an ideal condition, that is sometimes called 'local causes'. It is the requirement that each physical event or change has a physical cause, and that this cause can be localized in the immediate spacetime neighborhood of its effects. A collision of two billiard balls, or the mechanical connections between the parts of a steam engine are clear examples. A more subtle example is the feature of classical electromagnetism that any change in the velocity of a moving charged particle can be regarded as being caused by the action upon this particle of the electric and magnetic fields existing in the immediate spacetime neighborhood of that particle at the moment at which the change in velocity occurs, and that any change in the electric and magnetic fields are likewise caused by physically describable properties that are located very close to where that change occurs.

This idea that all physical effects are consequences of essentially 'contact' interactions was part of the intellectual milieu, stemming from the ideas of Rene Descartes, in which Isaac Newton worked while creating the foundations of modern physics. However, his universal law of gravitational attraction was stated as a law of instantaneous action over astronomical distances, a clear violation of the idea that all physical effects have local causes. Newton tried unsuccessfully to devise some local mechanical idea of how gravity worked, but in the end asserted his famous 'hypothesis non fingo' [I feign (pretend to make) no hypothesis (about how gravity works)] (Newton 1964/1687, p. 671). He relied, instead, on the empirical success of his simple inverse-square-law postulate to account for a huge amount of empirical data. Yet as regards the basic metaphysics he wrote (Newton 1964, p. 636):

> That one body can act upon another at a distance through the vacuum, without the mediation of anything else, by and through which their action and force my be conveyed from one to another, is to me so great an absurdity that I believe that

> no man who has in philosophical matters a competent faculty of thinking can ever fall into it.

This statement is a trenchant formulation of the notion of locality. It took more than two centuries of development before Einstein came up with an explanation, in terms of the idea of distortions of spacetime, that allowed the requirement of locality to be met for gravity. Einstein's special theory of relativity imposes the 'locality' condition that no localized measurable output can depend upon the character of a localized experimenter-controlled physical input before a point moving at the speed of light can travel from the smallest region in which the input is localized to the smallest region in which the output is located. This locality condition is required to hold in any physical theory that is called 'relativistic'.

This idea of locality is fairly simple and straightforward in classical physics, because in that setting everything has a material basis and all causal effect are associated with transfers of momentum or energy, which moves about in a continuous no-faster-than-light way. In quantum theory the fundamental substrate of causation is more ephemeral: causation is carried by *potentialities for observational events to occur*. These potentialities *usually* change in a localized continuous way, but, in conventional quantum mechanics, they change abruptly in association with the occurrence of an actual observation or observer-controlled input. And a 'cause', such as the performance of a freely chosen measurement procedure in one region, can have a certain kind of instantaneous faraway effect without any energy or momentum *traveling* from the region of the cause to the region of the effect.

In the quantum context a possible definition of locality pertains to *information*: it requires that no information about which measurement is freely chosen and performed in one spacetime region can be present in another spacetime region unless a point traveling at the speed of light or less can get from the first region to the second. Or in terms of outcomes: no statement whose truth is determined solely by which outcomes appear in one spacetime, under conditions freely chosen in that region, can be true if one experiment is freely performed in a region that is spacelike-separated from the first region, but be false if another experiment is freely chosen there. The term 'freely chosen' means only that in the argumentation this choice is not to be constrained in any particular way: that we are dealing with predictions that do not depend upon how the choice of measurement is determined or specified, except that it be designed to be physically independent of the system upon which the measurement is being performed. Locality defined in either of these latter two ways is violated in quantum theory.

F Einstein Locality and Spooky Action at a Distance

In 1935 Albert Einstein, in collaboration with Boris Podolsky and Nathan Rosen, published a landmark paper entitled *Can Quantum Mechanical Description of Physical Reality Be Considered Complete?* (Einstein 1935). Einstein had already been engaged for several years in a discussion with Niels Bohr about the completeness of quantum theory. In the 1935 paper Einstein did not challenge the claim of the quantum theorists that their theory was complete in the pragmatic/epistemological sense that it gives all possible empirically testable predictions about connections between the various aspects of 'our knowledge'. In the 1935 paper Einstein et al. effectively accepted this claim of epistemological completeness, but defined the question they were addressing to be the completeness of quantum mechanics *as a description of physical reality*.

'Physical reality' is a slippery concept for scientists, when it becomes separated from empirically testable predictions. Hence Einstein and his colleagues were faced with the difficult task of introducing this term into the discussion in a way that could not easily be dismissed as vague metaphysics by a physics community which, greatly impressed by the empirical successes of quantum mechanics, was in no mood to be sucked into abstruse philosophical dialectics. Yet Einstein and his colleagues did succeed in coming up with a formulation that shook the complacency of physicists in a way that continues to reverberate to this day.

The key to their approach was to tie the needed characterization of physical reality to a peculiar *nonlocal* feature of the quantum mechanical treatment of two-particle systems.

The mathematical rules of quantum theory permit the generation of a state of two particles that has predicted properties that appear, at least at first sight, to violate a basic precept of the special theory of relativity, namely the exclusion of instantaneous (i.e., faster-than-light) action at a distance.

Quantum theory generally allows any one of several alternative possible measurements to be performed on a particle that lies in some experimental region R. The choice of the measurement to be performed in R is treated in quantum mechanics as a boundary condition that can be 'freely chosen' by the experimenter. According to the Copenhagen interpretation, performing the measurement is supposed to affect the particle being measured in a way such that the observed outcome specifies the measured property of the state of the particle *after* the measuring process is complete. But then if two alternative possible measurements are *mutually incompatible*, in the sense that either one or the other can be performed, but not both at the same time, then there is no logical reason why the particle should have at the same time well defined values of *both* of the two properties.

The mathematical structure of quantum theory does in fact involve various properties of a particle that cannot, within that theoretical structure, have simultaneously well defined values. Potential inconsistencies are evaded by claiming that any two such theoretically incompatible properties are also empirically incompatible, in the sense that they cannot be measured simultaneously. But Einstein et al. constructed an argument designed to show that the values of certain of these properties are, nevertheless, simultaneous elements of physical reality. Such a demonstration would render the quantum mechanical account incomplete, as a description of physical reality!

To bring 'physical reality' into the discussion, in conjunction with the question of completeness, Einstein et al. noted that the basic precepts of quantum theory ensure that there is a state (wave function) of two particles that has the following properties:

1. The two particles lie at the time of a measurement performed on particle 1, in two large regions that lie very far apart.
2. There is a pair of measurable properties, X_1 and P_1, which are the location and the momentum of particle 1, respectively, that are neither simultaneously representable nor simultaneously measurable; and also a pair of measurable properties, X_2 and P_2, of particle 2 that are, likewise, neither simultaneously representable nor simultaneously measurable.
3. The prepared state of the two particle system, before the measurement is performed on particle 1, is such that measuring the value of X_1 determines the value of X_2, whereas measuring the value of P_1 determines the value of P_2.

These properties entail that the experimenter in the region where the first particle lies can come to know either X_2 or P_2, depending upon which measurement he chooses to perform. This choice controls physical measuring actions that are confined to the region where particle 1 is located, and this region is very far from the region where particle 2 is located. Consequently, any physically real property of the faraway particle 2 should, according to the precepts of the theory of relativity, be left undisturbed by the nearby measurement process: the distance between the two regions can be made so great that the physical consequences of performing the measurement on particle 1 cannot reach the region where particle 2 is located without traveling superluminally: faster than the speed of light.

These considerations permit Einstein et al. to introduce 'physical reality' by means of their famous 'criterion of physical reality':

> If, without in any way disturbing a system, we can predict with certainty (i.e., with probability unity) the value of a physical property, then there exists an element of physical reality corresponding to this physical property.

If a measurement were to be performed in the region where particle 2 is located then the quantum theorist could argue that this measurement could disturb the particle, and hence there would be no reason why properties X_2 and P_2, should exist simultaneously. But the situation under consideration allows either of the two (simultaneously incompatible) properties of particle 2 to be determined (predicted with certainty) without anything at all being done in the region where that particle 2 is located, and hence, according to the ideas of the theory of relativity, 'without in any way disturbing that system'. Thus Einstein and his colleagues infer, on the basis of their criterion of physical reality, that *both* properties are physically real. However, these two properties cannot be represented simultaneously by any quantum mechanical wave function. Hence Einstein et al. "conclude that the quantum mechanical description of physical reality given by wave functions is not complete".

Anticipating an objection Einstein et al. complete their argument by saying:

> One could object to this conclusion on the grounds that our criterion of reality is not sufficiently restrictive. Indeed, one would not arrive at our conclusion if one insisted that two or more physical quantities can be regarded as simultaneous elements of reality *only when they can be simultaneously measured*

> *or predicted*. On this point of view, since either one or the other, but not both simultaneously, of the quantities P [here P_2] or Q [here X_2] can be predicted they are not simultaneously real. This makes the reality of P and Q depend upon which measurement is made of the first system, which does not disturb the second system in any way. No reasonable definition of reality can be expected to permit this.

If one examines the situation considered by Einstein et al. in the explicit formulation of relativistic quantum field theory given by Tomonaga (1946) and Schwinger (1951) one finds that the quantum state (wave function) of particle 2 after the measurement is performed on particle 1 depends not simply on which measurement is performed on particle 1, but jointly upon which measurement is performed and what its outcome is.

In a general context it is neither problematic nor surprising that what a person can *predict* should depend not only upon which measurement he performs, but also upon what he learns by experiencing the outcome of that experiment, and hence upon both which measurement is chosen and performed, and which outcome then appears.

In *classical* relativistic physics an *outcome* in one region can be *correlated* to an *outcome* in a faraway region – that is spacelike separated from the first – without their being any hint or suggestion of any faster-than-light transfer of information. Such correlations can arise from a common cause lying in the earlier (preparation) region from which each of the two later experimental regions can be reached by things traveling at the speed of light or less.

In relativistic quantum field theory, as in relativistic classical theory, merely performing the measurement action on particle 1 does not affect any measurable or predictable property of particle 2. In both the classical and quantum versions the subsequent *outcome* pertaining to particle 1 is *correlated* (through the earlier initial preparation) to a predictable and measurable outcome pertaining to the faraway particle 2. Thus, although this experimenter's choice and his consequent action on particle 1 have, by themselves, no *direct* faraway effects, this choice and consequent action – by determining the physical significance (X_1 or P_1) of the local outcome, and thereby also the physical significance (X_2 or P_2) of the correlated faraway outcome – do influence the *nature* of the particular property of the faraway property of particle 2 that will be revealed to the experimenter who is performing the measurement on particle 1, by his experiencing the outcome of the experiment that he has chosen and performed. But this sort of 'influence' would, as in the classical case, fall far short of any indication of the need for any superluminal action at a distance, or of any superluminal transfer of information about the nearby free choice to the faraway region. All that has happened, in both the classical and quantum cases, is that the nearby experimenter has learned the value of an outcome that is *correlated* to the value of the outcome that a particular faraway experiment would have if the faraway experimenter were to choose to perform that particular experiment.

To identify what makes the quantum case different from classical case, suppose one has two balls, one red and one green, and one hot the other cold. Suppose they are shot in opposite directions into two far-apart labs. Simply measuring the color of the ball reaching the first lab does not immediately disturb in any way anything in the other lab. But knowing the outcome of this color measurement allows one to know something about what will be found if color is measured also in the second lab. *But in the classical case this real property of the system that arrives in the second lab would not be nullified or eradicated if one had chosen to measure temperature instead of color*. It is the claimed *nullification* of one *kind* of property of particle 2 or another, on the basis of which kind of experiment is performed on particle 1, that distinguishes the quantum case from the classical one. It entails the need for some sort of leaping of the information about which action was chosen and performed on particle 1 to the region where particle 2 is being measured. The need for this nullification arises from the fact that no wave function can represent a well-defined value of both X_2 and P_2.

In spite of this apparent violation of the notion that no information about the free choice made in region 1 can get to region 2, relativistic quantum field theory is compatible with the basic requirement of relativity theory that no 'signal' can be transmitted faster than light. A *signal* is a carrier of information that allows a receiving observer to know which action was taken by a distant sender. Because the receiver does not know, superluminally, which *outcome* was observed by the sender, she, the receiver, cannot know, superluminally, which action was taken by the sender. Hence no signal can be sent.

The sender, who knows both which experiment he has freely chosen and performed, and which outcome has appeared, knows, on the basis of his knowledge of both the theory and this outcome, more about what the receiver will experience than the receiver herself can know.

Quantum theory, by focusing on knowledge and prediction, is able neatly to sort out these observer dependent features. The theory carries one step further Einstein's idea that science needs to focus on what actual observers can know and deduce on the basis of their own observations. But quantum theory places a crucial restriction on definability that classical relativistic theory lacks: a person by his choice of probing action performed in one region can cause one *type* of property in a far away region to become *undefined in principle*, within the theory, because an incompatible *type* of property becomes defined there.

In the book entitled *Albert Einstein: Philosopher–Physicist*, Einstein (1951, p. 85) gives a short statement of his locality condition:

> The real factual situation of the system S_2 is independent of what is done with the system S_1, which is spatially separated from S_2.

The problem of reconciling this condition with quantum theory is that quantum theory is a theory of predictions (about outcomes of observations) not a theory of reality. The probing action performed on system S_1 by the experimenter does not, by itself, *disturb* in any way the real factual system S_2. This action, by itself, does not allow any new prediction to be made about any outcome of any measurement made on S_2. Hence one may quite reasonably claim that "the real factual situation of the system S_2" is not disturbed by the mere action of performing the faraway measurement. Yet it is in no way surprising that what kind of predictions one can make about the faraway correlated system – once the outcomes of the nearby measurement becomes known – depends upon what kind of nearby measurement is chosen. Einstein's challenge is to the following quantum theoretical claim: if the quantum state, which pertains to predictions, allows no predictions about a property then that property is *in reality* ill-defined.

If one accepts the quantum claim that the property itself is ill-defined if the property is not defined in the quantum theoretic state then the argument of Einstein et al. shows that the condition of no-faster-than-light action is violated in quantum theory. It is violated because the mere choice made in one region determines, no matter which outcome occurs, which *kind* of property of the faraway particle becomes, within the quantum framework, ill-defined.

The conclusion is that Einstein's argument leads, *within the quantum theoretical framework itself*, not to a proof of some incompleteness of quantum theory, but rather to a proof of the existence *within that theory* of a faster-than-light transfer to a faraway region of the information about which measurement is performed in the nearby region.

This conclusion depends, however, on accepting the basic precept of quantum theory that if two properties of a system cannot be simultaneously represented by a wave function and one of these two properties is defined then the other cannot exist. Einstein rejected that premise. The question thus arises: Can the requirement of no superluminal transfer of information be upheld if one relaxes the quantum precept, call it QP, that properties that cannot be simultaneously represented by any quantum state *cannot* be considered to be simultaneously definite.

This question was studied first by John Bell (1964) and later by others, within the special context of theories that postulate the existence of *hidden variables* that determine, simultaneously, the outcomes of all of the alternative possible measurements between which the experimenters are considered free to choose. Those arguments show that, within this hidden-variable context, the answer to the question posed at the end of the preceding paragraph is 'No'! Once the notion is accepted that decisions as to which measurements are performed are controlled by free choices that can go either way, it is impossible to reconcile *the predictions* of quantum theory for all of the then-allowed alternative possible measurements with the locality demand that the information about which measurement is freely chosen in a region cannot be present in any region that cannot be reached from the first without traveling faster than light.

But this proof of the need for faster-than-light transfer of information, replace the strong assumption QP by another strong assumption, the existence of 'hidden' variables that specify definite outcomes of all the possible measurements, whether they are performed or not. The question thus arises: Can the strong assumption of either QP or hidden variables be replaced by a more satisfactory assumption?

G Nonlocality in the Quantum World

In quantum mechanics the term 'nonlocality' refers to the failure of a certain relativity-theory-based locality assumption. This assumption is that no *information* about which experiment is freely chosen and performed in one spacetime region can be present in a second spacetime region unless a point traveling at the speed of light (or less) can reach some point in the second region from some point in the first. This assumption is valid in relativistic *classical* physics. Yet quantum theory permits the existence of certain experimental situations in which this information-based locality assumption fails

The simplest of the experiments pertinent to this issue involve two measurements performed in two spacetime regions that lie so far apart that nothing traveling at the speed of light or less can pass from either of these two regions to the other. The experimental arrangements are such that an experimenter in each region – or perhaps some device that he has set up – is able to choose between two alternative possible measurements. The locality assumption then demands, for each region, that the truth of statements the truth or falsity of which is determined exclusively by the outcomes of the possible measurements performed in that region be independent of which experiment is 'freely chosen' in the other (faraway) region.

The first actual experiment exhibiting these features was carried out by Aspect, Grangier, and Roger (Aspect 1981). Dozens of other such experiments have been carried out since, and the validity of all of the tested quantum predictions appear to be borne out. I shall accept the premise that all of the predictions of quantum theory pertaining to experiments of this kind are valid, even though some of them are yet to be performed.

The significance of this information-based nonlocality property of quantum theory is clouded by several considerations. The first is that this faster-than-light effect cannot be used to send a superluminal *signal*: no one can use this effect to transfer, superluminally, of information that he or she possesses to a faraway colleague. This limitation on *signal* velocity – together with other relativistic features of the theory – allows relativistic quantum field theory to be called 'relativistic' in spite of the entailed breakdown of this locality condition.

It might seem contradictory to assert first that locality fails, in the sense that information about which experiment is freely chosen and performed in a first region must be present in a second region, yet to assert, then, that the experimenter in the first region cannot use this feature to send information to a colleague in the second region. The resolution of the puzzle is that the dependence of faraway *measurable* properties upon the choice made by the nearby experimenter arises only via nature's choice of the *outcome* of the nearby experiment. The faraway colleague, lacking all knowledge about which outcome occurs in the sender's region, must treat that outcome as unknown. This leads to a quantum theoretical averaging over these outcomes that exactly eliminates all dependence upon the sender's free choice of anything that the receiving colleague can observe.

A second clouding consideration is this: in order to analyze the consequences of the non-dependence of some property upon a free choice one must consider, theoretically, or logically, within one argument, the consequences of various alternative choices. But, in the cases of interest, only one of the alternative possibilities can actually occur in any one existing empirical/experimental situation. Thus the argument needed to demonstrate the existence of faster-than-light transfer of information requires some sort of counterfactual reasoning that involves considering in one overall argument the quantum *predictions* about outcomes of several experiments that cannot all be actually performed.

A logical opening to counterfactual argumentation is provided by the precepts of quantum theory themselves. Bohr often emphasized the freedom of experimenters to choose which experiment is actually performed. This *freedom to choose* is important in quantum theory for the following reason: the quantum state (or wave function) of a physical system provides the information needed for a *prediction* about the outcome of *whichever* experiment is freely chosen and performed: predictions for all of the alternative possible choices of measurement are simultaneously imbedded in the quantum state, even though only one of the alternative possible measurements can be physically realized. Bohr's notion of complementarity rests on this aspect of quantum theory. The quantum claim of the pragmatic or epistemological completeness of quantum theory rests on the fact that there is an exact match between the sets of mutually compatible alternative possible outcomes that can be simultaneously represented mathematically by a set of quantum states and the sets of alternative possible outcomes that can be the empirically distinguishable outcomes of a experimenter's probing action. This intricate match between theory and empirical reality is the logical foundation of orthodox quantum mechanics.

The validity of this conceptual framework was brought into question by the 1935 paper by Einstein, Podolsky, and Rosen, discussed above. Because these authors were endeavoring to prove an internal inconsistency of the quantum precepts, they were careful *not to assume* that, contrary to the precepts of quantum theory, the outcomes of mutually incompatible measurements were simultaneously well defined. On the contrary, they used precisely the quantum prohibition on well defined values of mutually incompatible properties to deduce the conclusion that they could determine by their nearby choice of probing action which of two faraway mutually incompatible properties was defined. What they actually thereby proved was that Copenhagen precepts entailed the existence of faster-than-light transfer of information, though not the possibility of (relativity-theory-violating) faster-than-light signaling.

In 1964 John Bell published a follow-up to the 1935 paper of Einstein et al. Because it was, specifically, the Copenhagen prohibition against well defined values for the outcomes of mutually incompatible measurements that allowed Einstein et al. to deduce the need for faster-than-light transfer of information, Bell (1964) inquired whether dropping that Copenhagen precept could extinguish the need for faster-than-light information transfer. Bell forthrightly contravened the Copenhagen ban on determinate outcomes of mutually incompatible measurements by introducing 'deterministic hidden variables'. These hidden variables specify, simultaneously, the outcomes of *all* of the alternative possible experiments under consideration. Bell then showed that, within this deterministic hidden variable structure, one cannot reconcile the validity of the predictions of quantum theory (in these experiments) with the locality assumption that the outcomes in each region be independent of which experiment is performed in the other (faraway) region.

The hidden-variable machinery introduced by Bell is actually superfluous: all that is really needed is the assumption that in any given empirical *instance*, prior to the independent choices made by the experimenters in the two far-apart regions, any one of the four allowed pairs of choices *could* occur, and that for each such pair of choice (of which pair of measurement is performed) some long sequence of N pairs of numbers represent outcomes that *could* occur in the pair of regions if N repetitions of the selected pair of measurements were performed. The existence of such sequences of pairs of numbers specifying possible outcomes follows from Bell's hidden-variable machinery. But they refer only to *performable actions* and *observable outcomes*. Thus they can be stated without bringing in any notions of 'microscopic', 'invisible', or other 'hidden' variables. The assumption that such a set of pairs of numbers specifying possible outcomes exists is called 'counterfactual definiteness'. This assumption, *expressed at the macroscopic level*, cannot be consistently reconciled with the assumed validity of the predictions of quantum theory for each of the (four) measurement possibilities available to the experimenters, if one demands also that the outcomes in each region be independent of which experiment is chosen and performed in the faraway region (Stapp 1979).

Bell (1971) and others (Clauser 1969) went on to consider, instead of *deterministic* local hidden-variable theories, rather *probabilistic* local hidden variable theories. But, as shown by Stapp (1978), and independently by Fine (1982), this change does not substantially change the situation, because the two detailed formulations are, from a logical point of view, essentially equivalent.

The locality assumption fails, therefore, under either of these two opposing conditions on outcomes: either the Copenhagen prohibition of well defined values of outcomes of mutually incompatible measurements, or the counterfactual definiteness assumption that for each of the four possible combinations of measurements available to the experimenters, some set numbers represents outcomes that *could* occur if that pair of measurement were to be selected by the experimenters

In both of these to cases some special conditions pertaining to outcomes are imposed. The question thus naturally arises whether locality fails also under the weaker assumptions that, for some selected experimental situation, the predictions of quantum theory are valid and the two choices (one made in each of two very far apart regions, and determining which measurement will be performed in that region) can be treated as two independent free variables.

The answer is affirmative!

Under experimental conditions described by Hardy (1993) there are again two far apart experimental spacetime regions, labeled R and L, and in each region an experimenter chooses between a first or second possible measurement and he observes and records there whether the first or second possible outcome of the *single measurement* that he has just performed actually occurs. In some specific frame of reference the spacetime region L will be earlier than the spacetime region R. Quantum theory makes four pertinent predictions. The first two predictions combine with the locality condition that "the outcome observed and recorded in the earlier spacetime region does not depend upon which measurement is chosen and performed later" to prove, under the condition that the *first* of the two alternative possible measurements is chosen in the earlier region, the *truth* of the following statement (Stapp 2003):

> SR: If performing the first measurement in the later region R gives the first of the two possible outcomes, then performing there, instead, the second measurement would (necessarily) give the first of the two possible outcomes of that second experiment.

The truth of SR under the condition that the first measurement is performed in the earlier region follows from the combination of the first two predictions of quantum theory in the Hardy case. They are:

1. If the first measurement is performed in the later region and the first possible outcome appears there, then the first possible outcome must have appeared in the earlier region.
2. If the second measurement is performed in the later region and the first possible outcome appeared in the earlier region, then the first possible outcome must appear in the later region.

Combining these two predictions with the assumption that changing the choice of which experiment is performed in the later region cannot affect what has already happened earlier in the faraway region entails the truth of SR.

The second two predictions hold under the condition that the second measurement is performed in the earlier region. They are:

3. If the first possible outcome appears in the earlier region and the first measurement is performed in the later region, then the first possible outcome will appear in the later region.
4. If the first possible outcome appears in the earlier region and the second measurement is performed in the later region, then the second possible outcome will sometimes occur in the later region

Quantum theory predicts that no matter which of the measurements under consideration is performed, each possible outcome will occur half the time. Thus the common premise of (3) and (4) is sometimes satisfied. Combining these two predictions with the assumption that changing the choice of which experiment was performed in the later region cannot affect what already happened in the earlier faraway region entails that SR sometimes fails: the assertion SR is false.

The fact that statement SR about outcomes of measurements performable in the later region R is true if the first possible measurement is chosen and performed in the earlier region L, but is false if the second possible measurement is chosen and performed in that earlier region means that information about which experiment is performed in the earlier region must be present in the later region. This conclusion contradicts the locality condition that *information* about which choice is freely made by an experimenter in one region cannot be present in a second region unless the second can be reached from the first by traveling no faster than light.

The failure of this locality condition absolutely precludes the possibility that the real world actually conforms to the precepts of classical physics. We do live in a quantum world in which far-apart aspects are linked in ways quite contrary to the mechanistic conception of nature postulated by classical mechanics. A beautiful, intricate, and rationally coherent mathematical machinery has been discovered that transforms the mechanistic mindless concepts of classical physics over to a highly tested, useful, and accurate mathematical picture of a nonlocal reality in which our streams of consciousness are naturally and efficaciously imbedded. It would seem that the quantum conception nature is, from the perspective of science, the appropriate physics foundation of any ostensibly deep inquiry into the details of the mind–matter connection, and hence into the nature of our own being.

References

Aspect, A., Grangier, P., Roger, G. (1981): Experimental tests of realistic local theories via Bell's theorem, Physical Review Letters **47**, 460–463

Bell, J.S. (1964): On the Einstein–Podolsky–Rosen paradox. Physics **1**, 195–200

Bell, J.S. (1971): Introduction to the hidden variable question. *Foundations of Quantum Mechanics*, Proceedings of the Enrico Fermi International School of Physics, Course II (Academic, New York) pp. 279–81

Bell, J.S. (1987): *Speakable and Unspeakable in Quantum Physics* (Cambridge University Press, Cambridge)

Bohm, D. (1952): A suggested interpretation of quantum theory in terms of hidden variables. Physical Review **85**, 166–179

Bohm, D.J. (1986): A new theory of the relationship of mind to matter. The Journal of the American Society for Psychical Research **80**, 113–135

Bohm, D.J. (1990): A new theory of the relationship of mind to matter. Philosophical Psychology **3**, 271–286

Bohm, D., Hiley, D.J. (1993): *The Undivided Universe* (Routledge, London and New York)

Bohr, N. (1934): *Atomic Theory and the Description of Nature* (Cambridge University Press, Cambridge)

Bohr, N. (1935): Can quantum mechanical description of physical reality be considered complete? Physical Review **48**, 696–702

Bohr, N. (1958): *Atomic Physics and Human Knowledge* (Wiley, New York)

Bohr, N. (1963): *Essays 1958/1962 on Atomic Physics and Human Knowledge* (Wiley, New York)

Bunge, M. (1967): The turning of the tide. In: *Quantum Theory and Reality*, M. Bunge (Ed.) (Springer, Heidelberg, Berlin, New York)

Chalmers, D.J. (1995): Response. In: *Explaining Consciousness. The Hard Problem*, J. Shear (Ed.) (Academic, Thorverton UK)

Chase, H.W. (1917): Consciousness and the Unconscious. Psychological Bulletin **14**, 7–11

Chomsky, N. (1958): A review of B.F. Skinner's Verbal Behavior. Language **35**, 26–58

Clauser, J.F., Horne, M.A., Shimony, A., Holt, R.A. (1969): Proposed experiment to test local hidden-variable theories, Physical Review Letters **23**, 880–884

Dennett, D.C. (1994): In: *Blackwell Companion to Philosophy of Mind*, S. Guttenberg (Ed.) (Blackwell, Oxford)

Dennett, D.C. (1991): *Consciousness Explained* (Little, Brown & Company, Oxford)

Eccles, J.C. (1990): A unitary hypothesis of mind–brain interaction in the cerebral cortex. Proceedings of the Royal Society of London B **240**, 433–451

Eccles, J.C. (1994): *How the Self Controls Its Brain* (Springer, Berlin, Heidelberg, New York)

Einstein, A., Podolsky, B., Rosen, N. (1935): Can quantum mechanical description of physical reality be considered complete? Physical Review **47**, 777–780

Einstein, A. (1951): Remarks to the essays appearing in this collected volume. In: *Albert Einstein: Philosopher–Physicist*, P.A. Schilpp (Ed.) (Tudor, New york)

Everett, H. (1957): 'Relative state' formulation of quantum mechanics. Rev. Mod. Phys. **29**, 454

Fine, A. (1982): Hidden variables, joint probabilities, and Bell inequalities, Physical Review Letters **48**, 291–295

Fogelson, A., Zucker, R. (1985): Presynaptic calcium diffusion from various arrays of single channels: Implications for transmitter release and synaptic facilitation, Biophysical Journal **48**, 1003–1017

Gazzaniga, M. (2005): *The Ethical Brain* (Dana Press, New York)

Ghirardi, G.C., Rimini, A., Weber, T. (1986): Unified dynamics for microscopic and macroscopic systems. Physical Review D **34**, 470

Haag, R. (1996): *Local Quantum Physics* (Springer, Berlin, Heidelberg, New York) pp. 313–322

Hagen, S., Hameroff, S., Tuszynski, J. (2002): Quantum computation in brain microtules: Decoherence and biological feasibility. Physical Review E **65**, 061901-1–061901-11

Hameroff, S., Penrose, R. (1996): Orchestrated reduction of quantum coherence in brain microtubules: A model for consciousness, Journal of Consciousness Studies **3**, 36–53

Hardy, J. (1993): Nonlocality for two particles without inequalities for almost all entangled states, Physical Review Letters **71**, 1665–1668

Heisenberg, W. (1958a): The representation of Nature in contemporary physics, Daedalus **87** (summer), 95–108 (p. 100)

Heisenberg, W. (1958b): *Physics and Philosophy* (Harper, New York)

Hendry, J. (1984): *The Creation of Quantum Theory and the Bohr–Pauli Dialog* (Reidel, Dordrecht, Boston)

James, W. (1890): *The Principles of Psychology*, Vol. I (Dover, New York)

James, W. (1892): *Psychology: The Briefer Course*. In: *William James: Writings 1879–1899* [Library of America (1992), New York]

James, W. (1911): *Some Problems in Philosophy*, Chap. X, Novelty and the Infinite – The Conceptual View, *Writings 1902–1910* (Library of America, New York) p. 1061

Jarrett, J.P. (1987): Bell's theorem: A guide to the implications. In: *Philosophical Consequences of Quantum Theory*, J.T. Cushing and E. McMullin (Eds.) (Notre Dame U.P., Notre Dame) pp. 60–79

Joos, E. (1996): Introduction. In: *Decoherence and the Appearance of a Classical World in Quantum Theory*, D. Giulini et al. (Eds.) (Springer, Berlin, Heidelberg, New York)

Mermin, N.D. (1987): Quantum mysteries for anyone. In: *Philosophical Consequences of Quantum Theory*, J.T. Cushing and E. McMullin (Eds.) (Notre Dame U.P., Notre Dame) pp. 49–59

Misra, B., Sudarshan, E.C.G. (1977): The Zeno paradox in quantum theory, Journal of Mathematical Physics **18**, 756–763

Nahmias, E.A. (2002): Verbal reports of the contentents of consciousness: Reconsidering the introspectionist methodology.
`http://psyche.cs.monash.edu.au/v8/psyche-8-21-nahmias.html`

Newton, I. (1964/1687): *Principia Mathematica*, F. Cajori (Ed.) (University of California Press, Berkeley)

Newton, I. (1704): *Optics*

Ochsner, K.N., Bunge, S.A., Gross, J.J., Gabrieli, J.D. (2002): Rethinking feelings: An fMRI study of the cognitive regulation of emotion, Journal of Cognitive Neuroscience **14**, 8, 1215–1229

Pashler, H. (1998): *The Psychology of Attention* (MIT Press, Cambridge MA)

Pearle, P. (2005): Quasirelativistic quasilocal finite wave function collapse model, Physical Review A **71**, 032101.
`http://arxiv.org/abs/quant-ph/0502069`

Penrose, R. (1989): *The Emperor's New Mind* (Oxford, New York)

Penrose, R. (1994): *Shadows of the Mind* (Oxford, New York)

Putnam, H. (1994): Review of Roger Penrose, *Shadows of the Mind*, *New York Times Book Review*, 20 November, p. 7. Reprinted in AMS bulletin:
`www.ams.org/journals/bull/pre-1996data/199507/199507015.tex.html`

Rédei, M., Stöltzner, M. (2001): *John von Neumann and the Foundations of Quantum Physics* (Kluwer, The Netherlands) pp. 97–134

Schwartz, J., Begley, S. (2002): *The Mind and the Brain: Neuroplasticity and the Power of Mental Force* (Harper, New York)

Schwartz, J., Stapp, H., Beauregard, M. (2003): The volitional influence of the mind on the brain, with special reference to emotional self regulation. In: M. Beauregard (Ed.), *Consciousness, Emotional Self-Regulation and the Brain* (Advances in Consciousness Research Series, John Benjamins, Amsterdam, New York)

Schwartz J.M., Stapp H.P., Beauregard M. (2005): Quantum theory in neuroscience and psychology: A neurophysical model of mind/brain interaction. Philosophical Transactions of the Royal Society B **360**, 1309–1327. `http://www-physics.lbl.gov/~stapp/stappfiles.html`

Shimony, A. (1965): Quantum physics and the philosophy of Whitehead. In: *Philosophy in America*, Max Black (Ed.) (George Allen and Unwin, London)

Shimony, A. (1987): Our world view and microphysics. In: *Philosophical Consequences of Quantum Theory*, J.T. Cushing and E. McMullin (Eds.) (Notre Dame University Press, Notre Dame) pp. 25–37

Shimony, A. (1993): *Natural Science and Metaphysics*, Vol. 2, *Search for a Naturalistic World View* (Cambridge University Press, Cambridge)

Shimony, A. (1997): In: Roger Penrose's *The Large, the Small, and the Human Mind* (Cambridge University Press, Cambridge)

Stapp, H.P. (1971): S-matrix interpretation of quantum theory, Physical Review D **3**, 1303–1320

Stapp, H.P. (1972): The Copenhagen Interpretation, American Journal of Physics **40**, 1098–1116. Also in Stapp 1993/2004

Stapp, H.P. (1977): Theory of reality, Foundations of Physics **7**, 313–323

Stapp, H.P. (1978): Non-local character of quantum mechanics, Epistemological Letters, June 1978 (Association F. Gonseth, Case Postal 1081, Bienne, Switzerland)

Stapp, H.P. (1979): Whiteheadian approach to quantum theory and generalized Bell's theorem, Foundations of Physics **9**, 1–25

Stapp, H.P. (1993): *Mind, Matter, and Quantum Mechanics*, 1st edn. (Springer, Berlin, Heidelberg, New York)

Stapp, H.P. (1997): Science of consciousness and the hard problem, The Journal of Mind and Behavior **18** (2–3), 171–194

Stapp, H.P. (1999): Attention, intention, and will in quantum physics, Journal of Consciousness Studies **6**, 143–164

Stapp, H.P. (2001): Quantum theory and the role of mind in nature, Foundations of Physics **31**, 1465–1499

Stapp, H.P. (2002): The basis problem in many-worlds theories, Canadian Journal of Physics **80**, 1043–1052

Stapp, H.P. (2004a): *Mind, Matter, and Quantum Mechanics*, 2nd edn., Sect. 12 (Springer, Berlin, Heidelberg, New York)

Stapp, H.P. (2004b): A Bell-type theorem without hidden variables. American Journal of Physics **72**, 30–33

Stapp, H.P. (2005): Quantum interactive dualism: An alternative to materialism, Journal of Consciousness Studies **12**, 43–58

Stapp, H.P. (2006a): *Quantum Interactive Dualism II: The Libet and Einstein–Podolsky–Rosen Causal Anomalies* (Erkenntnis)

Stapp, H.P. (2006b): Quantum approaches to consciousness. In: *Cambridge Handbook of Consciousness*, M. Moskovitch and P. Zelazo (Eds.). `http://www-physics.lbl.gov/ stapp/stappfiles.html`

Stapp, H.P. (2007a): Quantum mechanical theories of consciousness. In: *Blackwell Companion to Consciousness*, M. Velmans and S. Schneider (Eds.)

Stapp, H.P. (2007b): Whitehead, James, and quantum theory. Mind and Matter

Schwinger, J. (1951): Theory of quantized fields I. Physical Review **82**, 914–927

Tegmark, M. (2000): Importance of quantum decoherence in brain process, Physical Review E **61**, 4194–4206

Titchner, E.B. (1898): *An Outline of Psychology* (Macmillan, New York)

Tomonaga, S. (1946): On a relativistically invariant formulation of the quantum theory of wave fields, Progress of Theoretical Physics **1**, 27–42

Velmans, M. (2000): *Understanding Consciousness* (Routledge, London)

Velmans, M. (2002): How could conscious experiences affect brains? Journal of Consciousness Studies **9–11**, 1–29

Von Neumann, J. (1932): *Mathematische Grundlagen der Quantnmechanik* (Springer, Berlin). Translated as: *Mathematical Foundations of Quantum Mechanics* (Princeton University Press, Princeton, 1955)

Von Neumann, J. (1955/1932): *Mathematical Foundations of Quantum Mechanics* (Princeton University Press, Princeton) Chap. VI

Watson, J.B. (1913): Psychology as the behaviorist views it. Psychology Review **20**, 158–177

Whitehead A.N. (1978): *Process and Reality*, corrected edition by D.R. Griffin and D.W. Sherburne (Free Press, New York). Originally published in 1929

Wigner, E. (1961a): The probability of the existence of a self-reproducing unit. In: *The Logic of Personal Knowledge*, M. Polyani (Ed.) (Routledge and Paul, London)

Wigner, E. (1961b): Remarks on the mind–body problem. In: *The Scientist Speculates*, I.J. Good (Ed.) pp. 284–302 (Heinemann, London, 1961); (Basic Books, New York, 1962). Reprinted in: *Quantum Theory and Measurement*, J.A. Wheeler and W.H. Zurek (Eds.) pp. 168–181 (Princeton University Press, Princeton, 1983)

Wigner, E. (1963): The Problem of Measurement. American Journal of Physics **31**, 6–15. Reprinted in: *Quantum Theory and Measurement*, J.A. Wheeler and W.H. Zurek (Eds.) pp. 325–341 (Princeton University Press, Princeton, 1983)

Zeh, H.D. (1996): The program of decoherence: Ideas and concepts. In: *Decoherence and the Appearance of a Classical World in Quantum Theory*, D. Giulini et al. (Eds.) (Springer, Berlin, Heidelberg, New York)

Zurek, W.H. (2002): Decoherence and the transition from quantum to classical – revisited, Los Alamos Science, 27 November, 2–25. `arXiv/quant-ph/0306072`

Index

Printing: Krips bv, Meppel
Binding: Stürtz, Würzburg